Distributed Networked Operations

Distributed Networked Operations

✦

The Foundations of Network Centric Warfare

Jeff Cares

Alidade Press, Newport, Rhode Island

iUniverse, Inc.
New York Lincoln Shanghai

Distributed Networked Operations
The Foundations of Network Centric Warfare

iUniverse books may be ordered through booksellers or by contacting:

iUniverse
2021 Pine Lake Road, Suite 100
Lincoln, NE 68512
www.iuniverse.com
1-800-Authors (1-800-288-4677)

First Printing

ISBN-13: 978-0-595-37800-5 (pbk)
ISBN-13: 978-0-595-82178-5 (ebk)
ISBN-10: 0-595-37800-5 (pbk)
ISBN-10: 0-595-82178-2 (ebk)

Printed in the United States of America

For Betty

Contents

Preface

This book is intended to describe with analytical rigor the concept of distributed networked operations, which is a refinement of what have popularly been called "network centric operations." Distributed networked operations envision combat being conducted by large numbers of diverse and small units—rather than by small numbers of generally homogenous, large units. Examples of the latter are sea battles between blue water fleets and strategic bombing campaigns.

In theory and to a significant extent in practice in Afghanistan and Iraq, distributed networked operations involve a mixed bag of naval, ground and air units, none of which is individually as powerful as a fleet, air wing or armored division. Operations of this sort do have advantages that more traditional approaches lack. For example, properly executed distributed networked operations are less dependent on the survival of individual units and thus should be harder for an adversary to disrupt.

The key is how the activities of geographically dispersed and functionally diverse units are orchestrated, or controlled. This is obviously a complex matter that requires a thorough understanding of the concepts upon which distributed networked operations are based. This book contributes to this understanding by examining control theories and their implications for force structure.

The motivation behind this book arises largely out of dissatisfaction with the existing literature on Network Centric Warfare (NCW). Since this concept was introduced in the mid-1990s, no substantial analytical progress has been made on understanding how NCW should work. Like a new model of a car, the original NCW ideas must stand up to analytical scrutiny and real world use. To make matters worse, there seems to be intolerance among some of the authors of NCW literature of valid criticism. This intolerance has created *ipse dixit* assertions for NCW that are impeding open and healthy discussion of military transformation. Without careful dialogue, the defense community has given many NCW claims an intellectual pass. Almost ten years after the initial NCW claims were made, the existing literature and subsequent Pentagon briefings have not brought the defense community much closer to understanding the mechanisms of advantage in distributed networked operations.

The responsibility for progress may ultimately lie with those who have not challenged NCW precepts vigorously enough. This book is a public attempt to initiate this challenge. I am well aware, however, that presenting and defending a new formalism might lead to similar *ipse dixit* defenses and self-referential proofs. For that reason, a website has been developed for public discussion on the content of this book, http://www.dnobook.com. The reader is invited to this website to provide critical comments about the contents of this book, participate in detailed discussion of distributed networked operations and offer alternative formalisms for understanding distributed networked operations.

It is a goal of this book to express how far the defense community still has to go in determining network centric value propositions. It is hoped that by presenting some rigorous yet rudimentary formalisms, this book will provide a good start on developing better foundations for Distributed Networked Operations.

Jeff Cares
Newport, Rhode Island
22 August 2005

Acknowledgments

Work on the theory of distributed networked operations began for me almost fifteen years ago when I was fortunate enough to have my thesis supervised by CAPT Wayne P. Hughes, Jr., USN (ret.) of the Naval Postgraduate School. He suggested a provocative idea, some simple Operations Research street wisdom that is even more true now than it was then: no simulation can make up for bad math—if the math you need doesn't exist, it's the analyst's job to develop it. Wayne has been a sage mentor and a dear friend. I have either sought his advice for or suspect his hand in many of the opportunities I have had in my career. I could never thank him enough.

If Wayne has been my analysis "sea-pappy", then CDR John Q. Dickmann, USN (ret.), has been a brother. So much of the contents of this book were developed with or inspired by him that he should probably claim co-authorship. He certainly deserves it. Dr. Ray Christian and Bob Manke of the Naval Undersea Warfare Center were also partners in the early formulation of these ideas.

Dr. Stu Kauffman and Dr. Jose Lobo of the Santa Fe Institute and Dr. Yaneer Bar-Yam of the New England Complex Systems Institute had an early and profound influence on this work. The theoretical underpinnings of distributed networked operations are pulled directly from their work. I benefited enormously from discussions with Jose and Yaneer, and I thank them for their extraordinary patience.

David Jarvis, Mike Bell, Jim Miskel and Dave Garvey at Alidade and Frank Tito at Swingbridge deserve significant credit for helping to move these ideas—and Alidade itself—to bigger and better places. Jim deserves special credit as editor, transforming my tortured technospeak into more readable prose. If the reader has difficulty with the text, however, it is most likely with those passages on which I obstinately ignored his excellent advice. Frank also deserves special thanks for formatting the book itself. Dr. Ralph Klingbeil of the Naval Undersea Warfare Center was the first to promote study of wolves as a metaphor for distributed networked operations.

As the endnotes attest, a great deal of this work was funded directly by Mr. Andy Marshall, Director, Net Assessment, and VADM Art Cebrowski, USN (ret.), Director, Force Transformation, both of the Office of the Secretary of Defense. Alidade also kept its doors open and lights on thanks to the continued efforts of Bill Glenney at the CNO Strategic Studies Group (SSG).

Those previously mentioned are also in the *de facto* Alidade "support group." At the head of this list is the Director of the CNO Strategic Studies Group, ADM Jim Hogg, USN (ret.). I am grateful to Admiral Hogg for convincing me not to retire from active duty but to continue my service in Newport for four more years—if not for that opportunity, this book and Alidade would not be. Also at the head of the list is Chris Meyer, formerly head of the Cap Gemini Ernst & Young Center for Business Innovation, now Chief Executive of Monitor Networks. Chris and his A-team, particularly David McIntosh and John Jordan, welcomed John Dickmann and me into the world of Fortune 50 strategic planning and gave us a seat and a voice. John Hanley deserves credit for introducing those of us at the SSG to the Santa Fe Institute and the topic of complexity. Others who deserve special thanks are Shane Deichman of Joint Forces Command, J9, BG Jimmy Khoo of the Singapore Ministry of Defense, Mike Neely of NuTech Solutions, Henrik Friman of the Swedish Defense College, Col. Frans Osinga, Ph.D., Royal Netherlands Air Force, Scott Borg of Dartmouth College and Commander Al Elkins of the SSG. I would like to thank Rear Admirals Mike Lefever, Dave Gove, Jan Gaudio, Sonny Masso and Jeff Lemmons for their ongoing support, encouragement and guidance.

Dick O'Neill of the Highlands Group deserves special thanks for promoting Alidade as if it were his own. I owe a profound debt of gratitude to Peter and Diane Durand of Alphachimp Studios for putting Alidade's words into compelling pictures, as well as for their friendship. I am also grateful for the support from the Business Network of the Santa Fe Institute (thank you Susan Ballati and Ann Stagg). A special thanks as well to Shelley and Larry Kraman.

Thanks to all who have contributed to (or lurked on) the Alidade Online Discussion. To "The Good Father", LCDR Francis S. Mulcahy, USN, CEC, thanks for keeping the Alidade gang in good humor and please keep your head down.

1

Introduction

Throughout our history, humans have organized themselves with a variety of collective networks designed to achieve a wide range of social, political, commercial or martial tasks. Since the advent of enhanced information technologies in the 1990s, existing collectives and their fundamental patterns of interaction began changing in revolutionary ways. Whether it is a business professional conducting real-time collaboration with partners across the globe or youngsters pairing off against each other in virtual video games, no aspect of society seems immune from the impact of technologically-enabled networks.

The military is no different. With robust communication networks, commanders who were once forced to rely on either physical proximity or intermediate levels of hierarchy to pierce the fog of war can disperse their forces and coordinate their behavior in near real time to generate massed effects. These two elements—distributed forces and networked control—hold the promise of revolutionizing warfare at both the tactical and operational levels, thus transforming the strategic options retained by those who commit such forces to war.

Options imply choice; choice implies a value proposition. The greatest challenge to transforming the world's militaries from the Industrial Age to the Information Age has been the development of a clear notion of the value that distributed, networked forces bring to modern combat (compared, of course, to the value propositions of Industrial Age warfare). To be fair, this same question goes largely unanswered in the Information Age transformational efforts of other human pursuits, but often in these other domains market forces determine value. In the so-called New Economy, winners emerge and losers fade away. Information Age executives use balance sheets, sales projections and consumer surveys to refine their hypotheses about the market value of new business models. Serendipity, trial and error abound. But commercial incentives and market vagaries are inappropriate for the sobering realities of national security. Militaries are extraordinarily expensive, the potential for unnecessary human sacrifice is too great and liberty and freedom are far too dear to become objects for speculation.

Digging deeper into the topic of defense valuation, the researcher sees a more profound challenge—there is strong evidence that the value proposition behind legacy forces had some properties that value-seekers would find quite queer. It is well known, for example, that leaders at all echelons in the Soviet Red Army were forbidden to go on the offensive unless long-hand, *in situ* "correlation of forces" calculations showed that they enjoyed a 3:1 force advantage. Two extra divisions were mandated to a fight that may have been won with a much smaller force. From an Information Age valuation perspective, the key here is that the calculations included an implicit cost of information, a kind of "tax of ignorance" that was all too common in the Industrial Age. The Soviets massed and committed three times the force required because they did not understand how the complexities of battle would unfold. In short, they paid quite a price in men and materiel to use brute force to turn a complex world simple.

Such approaches still pervade contemporary military acquisition investment decisions. For example, it has taken the US Navy a long time to develop acquisition programs that appropriately address combat in contested littorals. For most of the Cold War, littoral requirements were considered "lesser-included cases" to open-ocean operations. Since the littoral fight was considered smaller in scale to blue water combat, large platforms received the biggest share of shipbuilding budgets. Large ships were assumed capable of prevailing in the littoral because they carried much more combat power in their big hulls than would be required to fight smaller coastal forces.

Recently, however, the US Navy recognized a greater potential for adversaries to deny the complex littoral to American warships.[1] Under littoral combat conditions, a large ship, lesser-included case procurement strategy—a high-cost, brute force Industrial Age solution—incurs a similar tax of ignorance because it delivers units of combat power that are much larger than required. This strategy leads to yet more perversion of the value proposition: smaller, cheaper units of combat power may actually out-perform larger, more expensive platforms.

To its credit, the US Navy has begun development of smaller platforms more suited to the complexities of littoral warfare than large-scale open-ocean warfare. Other services in the US and abroad are coming to similar conclusions about their own force structures and a host of new concepts for smaller, networked, manned and robotic systems are under development. The need for an Information Age value proposition for these new forces and platforms has never been greater.

THE WOLF PACK

Although it is impossible to determine the operational characteristics of complex, distributed networked forces that do not yet exist, it is possible to examine other disciplines for force design principles, potential experimental hypotheses and new ideas about the command, control and operation of distributed collectives. One of the most fruitful disciplines to examine is the study of wolf pack behavior. Animal behavior has long inspired traditional military operations and pack behavior itself was a metaphor for U-Boat tactics in World War II. Peering more deeply into wolf behavior, however, shows that wolf collectives are much more complex than the historical operations of German submarine teams and a thoughtful reading of history—and wolf behavior—suggests that the rudiments of distributed, networked forces were evident even in the Bay of Biscay.

At the most basic level, wolf packs are similar to military forces in that wolves must both hunt and attack their quarry just as military forces must search out and destroy opposing combatants. From a distributed networked forces perspective, these two functions reflect the dual prerequisite for adaptive behavior: exploration and exploitation. Indeed, wolf researchers insist that wolves are every bit as adaptive as another famously fit species, the shark. But whereas the shark's talent for killing solo is its great source of fitness, the wolf survives because the pack succeeds, and pack success is collective. Just as distributed, networked forces can be defeated in detail without collective support from the rest of the force, ostracized wolves die alone soon after they are banished from their pack.

Figure 1.1—Packs Exhibit Leadership, Organization and Communication

Packs, like military forces, have leadership, organization and communication. Once established as the strongest wolves of the pack, the "alpha" male and female assume leadership of the pack and establish the pecking order—the organization—for the rest of the group. Each wolf communicates its rank in this hierarchy with a system of yelps, facial expressions and tail attitudes. All wolves know how to interpret this language and are therewith informed, for example, when to hunt, where to rendezvous or when to sleep.

Although wolves lack the analytical capacities of humans, pack behavior is far from superficial and suggests some important principles for the development, operation and leadership of distributed networked forces. For example, a typical Information Age warfare assumption is that each unit in a force requires high bandwidth communications so all units can directly share information with a common database and common situational representations. Wolf communication, however, is very low bandwidth yet nonetheless quite rich. A simple signal, such as an alpha male's erect tail, can convey that a hunt has begun. This communication is usually local, however, observed not by all wolves but by a small clique within the pack of similar status to the alpha. This small group sends their own signals to lower caste wolves and soon all know that prey is near and the hunt is on. While pursuing a moose through deep snow, the alpha will lead the pack single file and continually update the pack through the hierarchy in this manner. This collective behavior raises a very useful question for developers of distributed, networked force. Since the electrical power and communications constraints are quite challenging when combat capability is distributed among smaller networked vehicles, shouldn't engineers and operators take inspiration from the wolf pack to design simple, low power, low bandwidth communications systems for collective behavior?

Another aspect of wolf pack behavior, the assignment of pack resources to different prey, similarly informs the development of distributed networked forces. Wolves are indiscriminate killers and will attack and eat anything from squirrels to moose (with the important exception that they only attack humans in legend and fairy tales). Wolf packs dynamically reconfigure to create different size collectives to attack different size prey. When attacking a squirrel or rabbit, for example, wolves will kill independently yet still travel together as a pack. As the pack encounters larger prey, the proportion of the pack contributing to the kill grows larger. A small deer, for example, might require the coordination of two or three wolves. An adult moose would require the complex collective efforts of the entire pack. A common assumption in many concepts for distributed, networked forces is that standard-sized platform collectives are always assigned to the same collective problem. Again the collective behavior of the wolf pack raises an important question for developers of distributed networked forces: should operating concepts explicitly contain the same multi-scale tasking approach as the wolf pack?

Figure 1.2—The Pack Attacks

An investigation of tactics for the attack itself also provides inspiration for the design and operation of distributed networked forces, particularly when the quarry is very large prey like deer or moose. In a winter moose attack, an alpha male or female leads the pack single file through heavy snow. If the pack is able to close on a moose, the moose will stand its ground and stare down the alpha wolf. If the alpha decides not to back down but to attack, as soon as the alpha lunges at the moose, each wolf executes its own unique and simple attack maneuver. One or two wolves will dig their fangs into the moose's hindquarters while another will dig into the fleshy nose, holding on to distract and tire the moose. Another wolf will try to sever the moose's jugular while others attack forequarters. The wolves will relentlessly attack from all directions while individually avoiding the moose's sharp hooves and heavy antlers. If one wolf falters or is shaken off, another will assume the abandoned task until the moose succumbs. The collective effect of these simple maneuvers is both more effective than if every wolf accomplished the same tasks without collaboration and more complex than a moose can cope with.

Implications for distributed networked force design and operations are obvious: lethal, complex collective behavior can self-synchronize from simple individual activities. In addition, diversity and variety are important, but so is coordination and cross-training. Many concepts for distributed network forces, however, continue to see collectives merely as large numbers of standard platforms completing identical tasks, all produced affordably with Industrial Age economies of scale. Another question for the design and operation of future military hardware flows naturally from this pack behavior. Can Information Age economies of scope create the same collective behavior in distributed networked forces as the individually simple yet collectively complex and lethal wolf pack tactics?

A final inspirational example from the wolf pack is perhaps the least obvious: it is quite difficult to study how wolves actually live, hunt and kill. This is not only because wolves are most active in the areas where humans are not. Wolf packs are difficult to study because of a trait they share with all complex systems—it is exceedingly difficult to discern causality in a complex system from observation alone. The number of different ways that the wolf pack can coordinate during a hunt or attack is infinite, and researchers must watch a very large number of attacks before the patterns become evident. An inherent characteristic of distributed networked forces is that their intent and composition—their emergent behaviors—can remain obscure until commanders are ready to attack an enemy. The question for engineers and operators becomes, Can we build sufficient diversity into a force and train for emergence so that the force's motives and operations remain ambiguous to competitors yet precisely evident to collaborators?

There are many more wolf pack behaviors that can motivate the development of distributed networked forces. The wolf pack and other collectives in nature make excellent role models for distributed networked operations and are stimulating research into new value propositions for future combat.

Transforming Operational Art

Researchers looking for new sources of value should also note the extent to which the environment affects natural collectives. The idea that the environment strongly impacts military forces during combat is obviously not new, but this book puts forth a stronger notion: that military forces must not be valued outside of the context of their use. This suggests that military hardware has no *inherent* value, a provocative point that underscores the importance of what the military profession calls "operational art" to distributed networked operations. Operational art is an axiomatic approach to understanding how military forces should be used in a campaign to achieve victory. A typical war college curriculum devotes a great deal of study to how a military force should be deployed to a geographical space by a specific time or at a particular rate to attack an enemy's "center of gravity," achieving "culminating points" which lead to victory.

These notions of campaigning, however, are infused with Industrial Age assumptions because they were developed for Industrial Age combat. Prototypical Industrial Age processes are characterized by limited communications, concentration of physical mass, centralized control and inflexible decision making, with information that is difficult to obtain and hard to share. Information Technology has enhanced communications capabilities, but the premium placed on concentration of force and centralized control remains. IT has wrought high order change on Industrial Age systems: computer networks and advanced communications mean that systems can physically disperse, distribute information widely, decentralize their cognitive capabilities and readily share information. IT types of networks were the first manifestation of this change but changes in operational art have lagged. Any valuation of future military forces will be deeply flawed unless it comports with a new operational art that takes advantage of new physical, information and cognitive networks that, like the wolf pack, achieve success through purposeful orchestration of complex, collective behavior. A goal of this book is to lay the foundations for a renewed discussion of Information Age operational art.

Distributed Networked Forces

Throughout the last decade, long-range planners in the world's armed services and defense industry have been proposing concepts to leverage Information Technology advances to solve many long-standing military problems. These problems include destruction of time-critical targets, search and surveillance, long-range power projection, access to contested littoral areas and force-wide information processing.

A useful thought experiment to explore how military hardware might change to address these challenges in the Information Age starts with considering a traditional warship. Naval vessels are large, manned platforms with weapons, radars, sonars and communications systems welded to the hull and the superstructure. A first approximation of future military hardware can been envisioned by mentally breaking a ship's welds and re-distributing the weapons, radars, sonars and communications gear between the ship and other, smaller vessels, which are perhaps unmanned. The US Navy is planning such an array of platforms for littoral warfare. This is an example of what is meant by a "distributed networked force."

Other services also see similar distributed forces in their future. For many years, the US Marines have been developing locally operated unmanned reconnaissance drones and recently began development of a concept for "Distributed Operations." The US Army has been investigating an alternative to large hierarchical forces designed to fight the Soviets. The new Army concept, the Future Combat System, consists of a new class of vehicles, wide-spread employment of terrestrial drones for weapons, reconnaissance and logistics, and new communications networks to enable collective action. The US Air Force has been striving to break their dependence on a classic Industrial Age planning process, the Air Tasking Order cycle, in favor of a more adaptive and distributed air power planning tool. Similarly, the US Department of Defense has been prototyping a new distributed adaptive logistics network.[3]

These concepts assume extraordinary benefits from networking, but a rudimentary definition of networked force structure and the advantages of networked military behavior have only recently been expressed.[4] The purpose of this book is to explore the basic concepts behind networked systems, expose previously unrecognized challenges and recommend fruitful areas of research for concept development, analysis, experimentation and testing.[5]

ORGANIZATION OF THE BOOK

Understanding the command, control and engineering of the transformational capabilities of distributed networked forces is a high priority for militaries throughout the world. This book seeks to provide operationally focused guidance on distributed networked forces to leaders in the defense community by summarizing recent efforts to solve complex control problems in other environments. The book will relate these insights to the battlefield in three steps. First, the physical, informational and competitive environments in which future forces operate will be explored and analyzed. Second, the arrangement and command of future forces will be examined with particular attention paid to issues of composition and control. Third, three main questions currently perplexing military innovators will be addressed:

- What are the defining characteristics of a distributed networked force?
- What should a distributed networked force be capable of?
- How should distributed networked forces be developed to exploit their full potential?

This introduction has touched upon the major themes to be addressed on the path to achieving its purpose. Chapter 2 is a bridge from this introductory material to the book's main elements. It reviews fundamental ideas from complex systems research that are central to understanding collective behavior and the valuation of distributed networked forces. The chapter also addresses the perceived failings of existing Information Age command and control concepts and sets the stage for subsequent chapters to examine operational environments, command and control, and force structure.

Chapter 3 describes a formal method for defining the complexity of an environment. It offers specific examples of complex environments and discusses how military operations are affected by environmental complexity. The chapter shows how scales of observation influence complex competition and discusses how complexity is manifest in different domains of warfare.

Chapter 4 describes the command and control competencies required to employ a distributed networked force for advantage in a complex environment. The basic principles of adaptive control are addressed and the operational logic of distributed networked operations are presented and discussed.

Chapter 5 presents an Information Age Combat Model and explains how the model is used to understand distributed networked operations.

To place the earlier chapters in perspective, Chapter 6 contains an operational vignette that specifies how the ideas behind distributed networked operations might be applied in a future conflict. The concluding chapter culminates the discussion of distributed networked operations by answering the three questions, above.

Three appendices are included after the conclusion. First, recognizing that an Information Age value proposition and complementary new models transcend the military domain, Appendix I describes the top ten non-military applications for distributed, networked operations. Second, since most of the mathematics contained in this book, the mathematics of complex networks, have only recently been developed, Appendix II contains a primer on this topic. Third, the major ideas in this book are drawn from a very diverse array of technical research. Although the audience for this volume—military officers and defense community professionals—is quite well educated, almost none of the research behind the ideas in the book is found in military educational programs. To encourage further investigation by interested readers, Appendix III is a bibliography that lists the top 100 reference documents for research into distributed networked operations.

[1] See http://www.csbaonline.org for a discussion of the anti-access threat. http://peoships.crane.navy.mil/lcs/program.htm describes the US Navy's programmatic response to these threats.

[2] The descriptions of wolf pack behavior and photo credit for the wolf attack come from the seminal study by L. David, Mech, *The Wolves of Isle Royale,* Fauna Series No. 7—1966, National Park Service, Department of the Interior, (US Government Printing Office, Washington, 1966). The study is available in its entirety online at http://www.cr.nps.gov/history/online_books/fauna7/fauna.htm. Dr. Ralph Klingbeil at the Naval Undersea Warfare Center, Newport Division, deserves the credit for first developing the wolf metaphor for distributed, adaptive systems.

[3] Cares, Jeffrey R. and CAPT Linda Lewandowski, USN, *Sense and Respond Logistics: The Logic of Demand Networks,* undated, unpublished US Government white paper, 2002.

[4] Cares, Jeffrey R., Raymond A. Christian and Robert C. Manke, *Fundamentals of Distributed, Networked Forces and the Engineering of Distributed Systems,* NUWC-NPT Technical Report 11,366, 9 May 2002.

[5] Throughout the rest of this book the word "network" will refer to graph theoretic (link and node) representations of systems, not necessarily to information technology (IT) network structures, unless otherwise noted.

2

Points of Departure

This book is about how to build and use an adaptive military force. Many in the defense community, however, confuse IT—and sensor-enable reaction (*reactive response*) with true adaptation. Adaptation is a type of learning, but not the kind of learning derived from merely collecting, storing or retrieving data. Concepts that call for remote sensor grids to geo-locate all militarily relevant pieces of metal on a battlefield and automatically assign them to remote weapons systems are examples of this type of learning-by-database. This approach to learning suggests that information is created by the arrival of sensors, leading to sequential, pre-programmed responses and is nothing more than rarified Industrial Age warfare. Adaptive response, by contrast, recognizes that all the information in the battlefield exists before the arrival of sensors and stresses that the C2 task is one of not just collecting but also translating information. Adaptive learning requires high rates of feedback from the environment to drive continuous change and modification. The information is translated by creating a *model* of the information, that is, a representation of the information that is more concise than the information itself. If this model is sufficiently valid, it creates the kind of learning that a system can use to both improve current system fitness as well as to anticipate future outcomes. The system does not wait for specific elements of information before it acts—it can perform quite well by acting on the anticipation itself. Adaptive response allows a system to respond smoothly to changes in information conditions, avoid "knee-jerk," reactive responses and more likely survive shocks to the environment.

This distinction between reactive response and adaptive response will be discussed in greater detail in Chapter 4. The idea is briefly introduced here because it embodies two fundamental aspects in which distributed networked systems are different from traditional systems. The first aspect is an approach to accounting for uncertainty in competitive environments and the second is a different philosophy for creating models of military systems. This chapter addresses both of these issues.

UNCERTAINTY

Driving uncertainty out of military systems is a common goal of engineers and operators. This goal is typically manifested in systems that are increasingly *precise*. An example of a precise system is one in which extraordinary effort is expended to decrease the target location error (TLE) of enemy targets so that a weapon with a small circular error probable (CEP)[1] can be applied, maximizing the effect of rounds on the target and minimizing wasted rounds and collateral damage. Precision is a goal of concepts like "Effects-Based Operations," which advocate providing the right information at the right time to the right decision-maker to apply the right weapon for the right effect.[2] Such approaches put extraordinarily tight constraints on a system that only a small subset of all available information can satisfy. Precision warfare focuses on reactive response and requires either simple information conditions, extraordinary investment in sensors or heroic search efforts. As we will see in Chapter 3, information conditions in competitive environments are far from simple, so in practice precision warfare either consumes extraordinary effort or system performance falls short of projections.

A useful way of approaching the information conditions in competitive environments is to think about a continuum of risk to uncertainty and complexity. An environment characterized by *risk* is one in which the probabilities of all possible outcomes are known. Decision makers examine the odds that certain states will result from different actions and determine a course of action based on risk-reward assessments. For example, if a course of action is 70-per cent likely to succeed (30-per cent likely to fail) then a good decision maker can decide to assume the risk and try for success. An *uncertain* environment is one in which it is possible to know the outcomes of all states. In these conditions, decision makers use common approaches like decreasing CEP and TLE to increase precision and drive out uncertainty. A *complex* environment is one in which trajectories to future states cannot all be known in advance because of strong dependencies among states. Good decision makers in complex environments recognize that it is a fool's errand to merely increase precision. They know that the command and control process should be directed at discovering the motives, dynamics and long-term evolutionary behaviors of competitors in the environment.[3] These decision makers try to attain more than a reactive response to information on the battlefield; they attempt to create a deeper mental model of the environment to anticipate how competition might unfold.

An important implication of this risk-uncertainty-complexity construct is that situational awareness in competitive environments depends on more than just the platforms or sensors employed to achieve situational awareness, but the information conditions as well. Many sensor-heavy future concepts, however, suggest that no information exists until sensors arrive and that better situational awareness, even in complex environments, is simply a matter of improving the acuity of sensors.

The Example of "Approximate English"

In Claude Shannon's famous 1948 paper, *A Mathematical Theory of Communication,* there is a thought experiment using the English language that makes these abstract issues more concrete and shows that merely reducing uncertainty cannot overcome complexity.[4] The example creates sentences of "approximate English" by selecting letters and spaces based on different statistical sampling rules. Rules are successively more sophisticated, resulting in increasingly specialized sentences.

The first sentence, a *zero-order* approximation of English, results from assuming that letters and spaces are independent from each other and all occur with the same frequency. In this version of approximate English, the letter *e*, for example, is just as likely to be selected as the letters *x*, *j* or *d*. The sentence that results is:

XFOML RXKHRJFFJUJ ZLPWCFWKCYJFFJEYVKCQSGHYD QPAAMKBZAACIBZLHJQD

This sentence has none of the structure of English (and is actually a zero-order approximation of *any* language that uses the same 26 Roman letters and a space). A *first-order* English sentence, by contrast, results from assuming that letters and spaces are still independent from each other but occur with the frequency found in Standard English text. In this version of approximate English, just like in the popular word game, Scrabble™, the letter *e* is more likely to be appear than the letters *x*, *j* or *d*. The sentence that results is:

OCRO HLI RGWR NMIELWIS EU LL NBNESEBYA TH EEI ALHENHTTPA OOBTTVANA-HBRL

Although this sentence again carries no meaning, an observer can begin to create a model of English from this sample. An observer who collects a large enough sample of characters would determine the same Scrabble™ letter frequencies and use them to create similar first-order sentences.

A more sophisticated English, a *second-order* approximation, has the same frequency of letters and spaces as first-order English, but the letters are no longer independent. They are presented with the interdependent digram (two letter) structures found in Standard English, that is, a space is more likely to follow an *s* or a *d* than, say, a *u*, and common digrams like *nt* and *ch* are evident. The sentence that results from this rule set is:

ON IE ANTSOUTINYS ARE T INCTORE ST BE S DEAMY ACHIN D ILONASIVE TUCOOWE AT TEASONARE FUSO TIZIN ANDY TOBE SEACE CTISBE

Note how actual words are starting to form, e.g, *on, be, at* and *are*. These are a result of statistical structure, however, and do not contain real meaning. The next sentence has the same letter and space frequencies but extends the interdependencies to *trigrams* such as *ist*, *ion* and *the*. This *third-order* approximation of English results in the sentence:

IN NO IST LAT WHEY CRATICT FROURE BIRS GROCID PONDENOME OF DEMONSTURES OF THE REPTAGIN IS REGOACTIONA OF CRE

The rudimentary patterns of larger words are starting to form because more structure is added, that is, the letter and space arrangements are less random. Although words like *demonstures* are technically gibberish, they are nonetheless almost recognizable as English. One might imagine an experiment in which students are given this word during a vocabulary test and asked to define it and use it in a sentence. Very few students would probably conclude the word was artificially constructed. This underscores, however, that reduced uncertainty about the pattern of letters does not necessarily equate to meaning. We know exactly what the letters are and could derive the rule set with a sufficiently large sample of similarly constructed sentences. The fact that the words have no meaning shows the limits of uncertainty-reducing methods to create meaning. On the other hand, that an English-speaking human might use her internal model of English to impose meaning on these words shows adaptive response in practice (although in this case, of course, the imposition is misguided).

Approximate English can be constructed from larger elements than letters and spaces. A *zero-order word* approximation would use words independently of each other and assume words like *a* and *the* just as likely as *sprocket* or *fingernail.* A *first-order word* approximation would preserve independence but present words with the same frequency that they are found in Standard English text. A sentence that results from this rule set is:

REPRESENTING AND SPEEDILY IS AN GOOD APT OR COME CAN DIFFERENT NATURAL HERE HE THE A IN CAME THE TO OF TO EXPERT GRAY COME TO FURNISHES THE LINE MESSAGE HAD BE THESE

This is clearly recognizable as English, but both grammar and meaning are missing. A *second-order word* approximation is an even more sophisticated artificial English. In this version, both word frequency and word transition probabilities (probability that a word follows another word) are correct but no further grammar is included. The sentence that results is:

THE HEAD AND IN FRONTAL ATTACK ON AN ENGLISH WRITER THAT THE CHARACTER OF THIS POINT IS THEREFORE ANOTHER METHOD FOR THE LETTERS THAT THE TIME OF WHO EVER TOLD THE PROBLEM FOR AN UNEXPECTED

The inclusion of martial words is purely accidental, yet they reinforce the application of this thought experiment to the military domain. If this were an order received over a military communications circuit, a decision maker might mistakenly assume it was garbled in transmission rather than an artificial sentence. Using an internal model of English, the decision maker might attempt to derive a superior's intent from the message. This behavior is actually quite common in during real operations when pressure is high and decision makers are flooded with information.

The risk-uncertainty-complexity continuum can be seen in approximate English. Knowing the rule set for the low-order approximations makes it fairly straightforward to determine meaning: the letters and words are random, so there is no deeper meaning. This is a case of risk: the probability that the next letter will be a *p* or the next word will be *on* or *good* in a first-order approximation can be calculated. An observer that does not know the rule set in a low-order approximation is dealing with uncertainty: a large sample of letters or words must be analyzed before a rule set can be devised. There is an upper limit on this effort, however: the best that the observer can do is to determine the probabilistic rule sets (that is, pursue higher degrees of precision). An observer that is looking for meaning is dealing with complexity: the precise letters that make up the writing are less important than understanding the unique blend of context, grammar and intention that created the message.[5] Indeed, the English language is about 50 per cent *entropic*, that is, roughly half the letters or words could be removed from a Standard English passage and it would still retain most of its meaning, so long as the reader had a well-developed model of English. This means that observers who pursue higher degrees of precision could be wasting about half their time on activities that do not contribute to the meaning of an English sentence.

INTERDEPENDENCE AND SKEW DISTRIBUTIONS

The dual properties of interdependence and skew distributions are important to pattern-formation in approximate English. Interdependence is an obvious property of digrams, trigrams, and Standard English grammar. Less obvious is the fact that these spelling and grammatical structural relationships can be portrayed as a network, a system of links and nodes, often with important feedback and feedforward mechanisms. These mechanisms result in "rich-get-richer" schemes that create skew distributions of, say word frequency (skew distributions describe frequencies of occurrence when a small number of elements have a high occurrence, a moderate number have a moderate occurrence and a large number have infrequent occurrence). For example, the requirement that English sentences have subjects and the fact that subjects are very often modified by articles (*the*, *a*, *an*) means that the frequency of articles is extremely high (even though the frequency of specific nouns used as subjects might be low). In a network representation of English, words could be nodes and grammar rules could be links. The rich-get-richer mechanism for articles comes from the very large population of nouns linked to only three articles. Since these noun-nodes are linked to other words by the rules of grammar (e.g., verbs can be nodes and subject-verb agreement can be a link), an extraordinary number of paths in this English network will feed through the articles. More sentences mean more articles are used.

One skew distribution in English is the Zipf distribution, which describes the fact that a few letters (such as *e* and *s*) are quite prevalent, a moderate number of letters (such as *m*, *r* and *d*) are somewhat less so, and a larger number (such as *f*, *q*, *x*, *z*, and *j*) are more rare.[6] Figure 2.1 shows the frequency of occurrence for letters, spaces and punctuation in the first 1000 characters of the first chapter of Mark Twain's *Following the Equator*. The horizontal axis shows the number of times a particular letter appears and the vertical axis shows how many letters appeared that many times (in intervals of 10). For example, nine letters appeared between 0 and 10 times, 3 appeared 50 to 60 times and 1 appeared between 180 and 190 times. Figure 2.2 shows the frequency of words in the first 1000 words of that same chapter. Note how the distributions are skewed to the left but there are high values in the "tails" to the right.

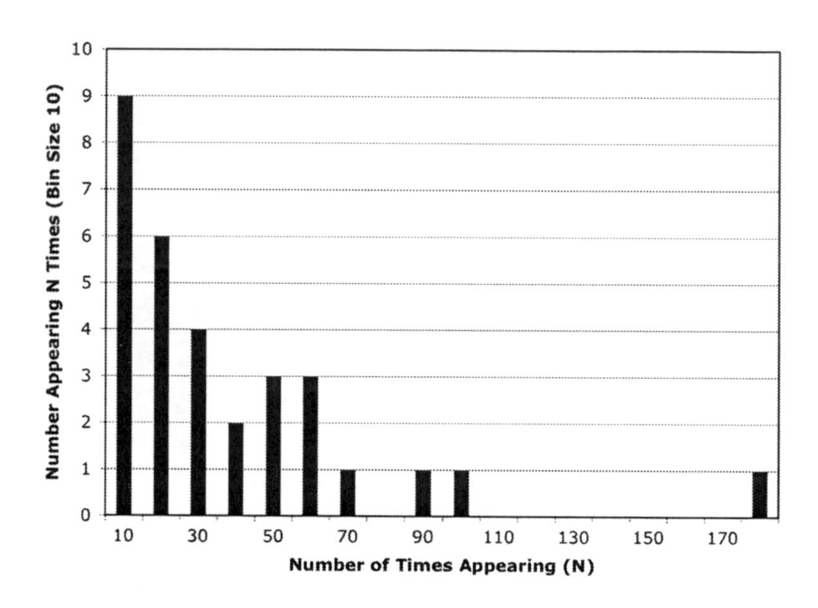

Figure 2.1—Frequency of Characters in Text

Despite the newness of IT-enabled military networks, interdependence has always been a characteristic of combat. Not surprisingly, then, skewed distributions are also found in combat performance measurements. Data on "ace-factors" and other skewed performance ratings go back many decades. For example, Figure 2.3 shows the distribution of kills in air-to-air combat in Vietnam for US Navy pilots (grouped by the aircraft carrier assignments for the period indicated).[7] Of the 61 total kills, 27 were achieved by pilots on only 2 of the aircraft carriers. In other words, almost half the kills were credited to pilots on 12 per cent of the carriers. Contemporary models of warfare processes, however, do not account for interdependence and skewed performance. This leads to the second point of departure: the need for a fundamental change in combat modeling.

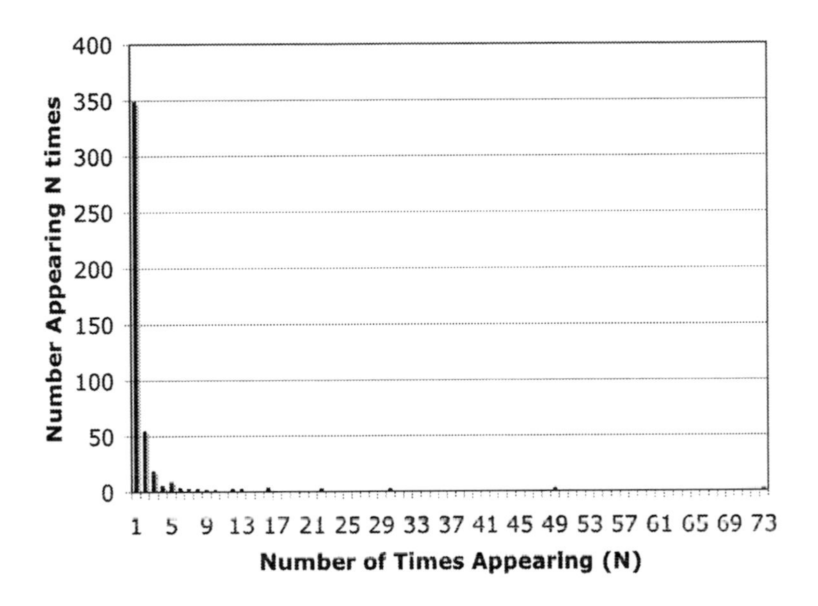

Figure 2.2—Frequency of Words in Text

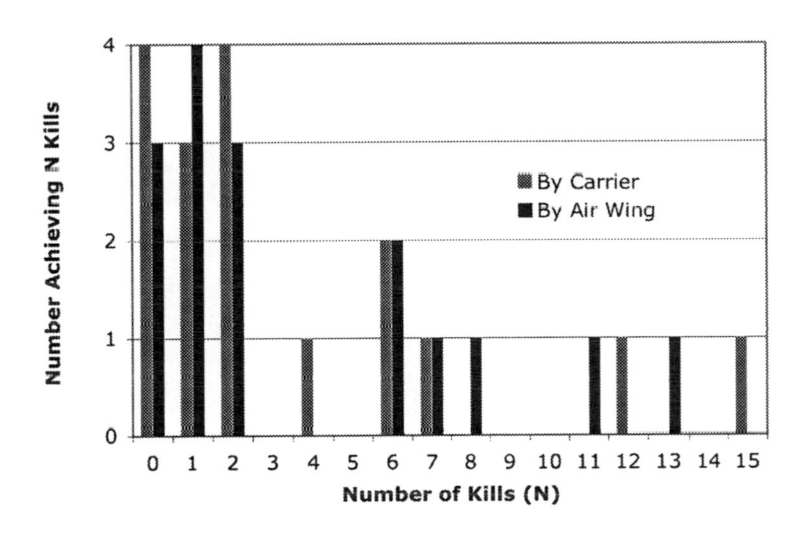

Figure 2.3—Frequency of Navy Air-to-Air Kills in Vietnam

COMBAT MODELING

A common mistake in many defense decision-making contexts is that "modeling" is conflated with "simulation." While an increasing number of operational and executive decisions depend on the results of a growing list of large, complex computerized renditions of combat, only a small fraction of analysts who use these simulations fully understand the mathematical relations, or models, that drive them. This can impart a false sense of formality and validity to the decisions the models support. Analysts endorse analytical results as having come "from the simulation," as if that alone constitutes analytical validity. The match between the mathematical guts of a simulation and structure of the simulated problem is often ignored. Despite widespread agreement in the importance of verification, validation and accreditation (VV&A) of simulations, VV&A is inconsistently applied in practice—particularly with regard to the suitability of mathematical models to represent real-world processes.[9] A new challenge has also arisen: mathematical models for distributed networked processes are not yet well developed. Analysts can only approximate new behaviors with old simulations, making the modeling-simulation conflation worse.

The following sections review mathematical models commonly used in warfare analyses and explore whether they have the potential to describe Information Age combat processes. The review will begin with a discussion of the assumptions underlying existing warfare models and the exploration will continue with an assessment of the analytical community's ability to describe Information Age warfare. Characteristics of a model that sufficiently describes and supports understanding of Information Age combat processes will also be defined.

TYPES OF COMBAT MODELS

Although there is great variety in the specific application of combat models, the fundamental structure of most combat models is one of two basic types: deterministic (closed-form) or stochastic (probability-based) combat models.

Deterministic Models. A deterministic model is one that fully describes all states in a given system with a set of closed-form equations. For example, Newton's Second Law, $f=ma$, which defines force, f, as the product of mass, m, and acceleration, a, does not admit for randomness or uncertainty in the mathematical representation of force. Because it completely "determines" theoretical relationships in a closed-form representation, the Second Law equation is a deterministic model.

Stochastic Models. Deterministic models were particularly useful before digital computers made automated calculations more routine for the analysis of large, intricate systems. Starting in the 1970s, however, the widespread use of digital simulation facilitated the inclusion of randomness and uncertainty in analytical models. A model that explicitly includes randomness is called a stochastic model. Stochastic models are commonly constructed by modifying one or more of the terms in a deterministic equation with random draws from an appropriate probability distribution. Simulations based on stochastic models must be run a certain number of times so that most of the randomness cancels out and results converge to stable statistical values. An important mathematical theorem, the Central Limit Theorem, computes how many times a stochastic simulation must be run. The most important factors in this calculation are the characteristics of the probability distributions in the model.

USEFUL COMBAT MODELS

It has been said that all models are wrong, but some models are useful.[10] All models must be wrong because they cannot represent the infinite detail of the real world. Modelers must therefore decide which details to include in a representation and which details to exclude. Included details are said to be *explicitly* modeled and the excluded detail is *implicitly* modeled. Representation decisions have an impact on usefulness: a model is *useful* if it captures the essence of real world processes sufficiently so that one understands the real world better after the model was constructed than before.[11] If too little detail is explicitly modeled, then an analyst's understanding of the real world process can suffer. If too much detail is explicitly modeled, then an analyst can have a virtual representation of the real world that is no closer to being understood than the real world was to begin with. Another way a model can be rendered useless is if the real world changes significantly. This is the case with many legacy models. Their math is fine; they just no longer apply to new combat processes. The underlying assumptions of legacy deterministic and stochastic models can indicate whether or not they are still useful.

Assumptions in Deterministic Models. Deterministic models are complete, closed form specifications of a system. Obviously, one cannot practically write an equation for every possible state or interaction that is observable in the process being modeled, so implicit representations necessarily abound. Two of the most prevalent implicit representations in deterministic models are aggregation and dis-aggregation. Modelers assume complex interactions at fine scales can be well represented at coarser scales by homogenizing behaviors and aggregating their effects. Aggregation relegates the differences in local behaviors throughout a system to a type of noise that can be represented by a parameter without loss of usefulness. Dis-aggregation is more challenging: it assumes coarser scale parameters can be reduced to pockets of differentiated local behavior. Aggregation and dis-aggregation assumptions apply to more than performance parameters: they assume that environmental conditions and localized tactics can be likewise aggregated and disaggregated.[12]

Another assumption prevalent in deterministic models is regularity. This assumption requires that small changes in input values must not trigger grossly non-linear outcomes. For example, doubling the value of an input should roughly equate to a doubling of its effect in the model and not, say, decreasing its effect by a factor of ten.

Finally, command, control or competitive processes are rarely explicitly represented in deterministic models. These effects are usually implied by such devices as the mathematical relationships between terms or the relative sequence in which processes are represented or invoked in computer code.[13]

Assumptions in Stochastic Models. Stochastic models also contain aggregation and regularity assumptions analogous to those found in deterministic models. Stochastic models have three additional significant assumptions. The first assumption stems from a basic problem in modeling randomness: since randomness implies incomplete knowledge of the input data required for a model, then some input parameters must be random variables. The second assumption flows from the first: since the data itself is to some extent random, so are the interactions between inputs. The parameters must therefore be treated as independent random variables. In other words, complex chains of causality in the operational processes being modeled are considered inconsequential. Interdependence is therefore misrepresented.[14] Third, modelers commonly assume that probability distributions for these independent random variables are not too skewed and the Central Limit Theorem will mandate a relatively small number of model runs.[15]

Command and control (C2) is more explicitly represented in stochastic models than in deterministic models. Stochastic models typically represent C2 processes in the same way other processes are represented, with draws from a probability distribution. A classic example of modeling C2 with stochastic techniques can be found in Anti-submarine Warfare (ASW) simulations. Initial detection of, say, a submarine by a surface ship's sonar is drawn from a probability distribution; so is the probability that the submarine transitions from detection to tracking. As the simulation proceeds, a random number is drawn at specified time steps to determine if the ship continues tracking the target or, alternatively, if the submarine becomes undetected, whereupon the initial detection probability distribution is invoked. Other C2 tasks, such as communication, radar detection, etc., are treated in a similar manner.

COMMON COMBAT MODELS

Some of the earliest deterministic attrition models described continuous fire in combat, where one side erodes the combat power of another at some fixed rate over time. The most prevalent example of a deterministic combat model are the eponymous Lanchester Equations, first published by a Victorian-era engineer who developed a mathematical force-on-force theory of combat in 1914.[16] In brief, Lanchester theorized that each side in a combat duel degrades the other side at some rate proportional to its own remaining size multiplied by the firing rate of its shooters. Using differential equations, Lanchester Equations prescribe such results as the ultimate winner of a contest between combatants, the time required for a duel to conclude or the size of each force remaining at a duel's conclusion. This model is the basis for most of the attrition-based combat simulations in use today. Indeed, dozens of variants of Lanchester's model are currently representing ground or air combat operations.[17] TACWAR is a theater-level deterministic ground and air combat simulation that employs Lanchester Equations as its underlying model of attrition.[18]

Two popular uses of stochastic models are for representing antisubmarine warfare (such as described in the previous section) and for air defense. The Naval Simulation System (NSS),[19] the Extended Air Defense Simulation (EADSIM)[20] and Joint Warfare System (JWARS)[21] are three prominent stochastic combat simulations that employ stochastic models.

SALVO MODELS

In the late 1980's Hughes brought combat modeling into the Missile Age by developing an attrition model inspired by the exchange of striking power during the Battle of Midway in World War II. His Salvo Exchange model describes combat as a pulse of offensive combat power designed to instantaneously penetrate an adversary's active defenses and cause damage to enemy platforms.[22] Although this model has important descriptive power, two major drawbacks to its predictive power are that it only holds for identical, homogeneous forces and it is strictly deterministic. Of course, homogenous force-on-force scenarios will be rare, and deterministic attrition obscures the importance of information or the sequencing of attacks.

The latter shortfalls were later addressed by introducing a version for heterogeneous forces as well as a stochastic variant.[23] The heterogeneous variant requires a high-dimensional "matching matrix" to define every interaction between elements of offensive combat power, defensive combat power, and staying power, but the problem of deterministic attrition remains. The stochastic version only models homogenous forces. Although these two variants were never combined into a stochastic, heterogeneous salvo model, such an exercise would be impractical. A full description of the stochastic matching matrix would be tantamount to an *a priori* description of all possible salvo exchange arrangements, obviating the need for the model to begin with. In other words, an analyst would have to create all the possible outcomes as an input to analysis rather than produce one as a result of analysis.

One powerful feature of the salvo model is its use for explicit calculations of *combat entropy*, a very normal condition of warfare. Combat entropy is one aspect of the "fog of war" in which assignment of combat power to targets will almost always be sub-optimal. These calculations have been used to explain the extent to which combat entropy and the sub-optimal assignment of combat power affects combat outcomes.[24]

SUITABILITY OF COMMON COMBAT MODELS TO DESCRIBE INFORMATION AGE COMBAT

Industrial Age combat was characterized by limited communications, massed forces, and centralized command, control and decision making. Information was difficult to obtain and hard to share, so commanders relied more on force than on a deep insight into tactical intricacies. Although important exceptions existed such as Fleet Anti-air Warfare, which pioneered rudimentary digital networks more than three decades ago, the more prevalent sentiment was exemplified by the Soviet's insistence on a 3:1 force advantage (as noted above, page 3).

Information technology and computer networks have been introduced into military processes to improve on this brute force approach by enabling the exchange of information for physical force where appropriate. As a result of these efforts, military systems are increasingly characterized by dispersal of physical assets, information distribution and decentralized cognition.

Since they have no better tools, defense analysts continue to use traditional models to simulate new, Information Age operational concepts. Often these models are embellished with additional C2 parameters (in the case of deterministic models) or the addition of C2 statistical terms (in the case of stochastic models). Newer efforts such as JWARS have attempted to more explicitly capture the most important command, control, communications, computers, intelligence, surveillance, and reconnaissance (C4ISR) operations. However, the underlying approach of models like JWARS has not departed from merely modifying traditional attrition models with C2 parameters or processes. Existing models have failed to fully represent the potential impact of Information Age command and control on combat outcomes because they are all based on physical models of attrition. In these physical models of attrition, the advantage accrued through greater numbers of assets or higher firing rates is more highly valued than advantage accrued through arrangement of assets. For this reason, none of the common models mentioned in the previous two sections are suitable candidates for an Information Age combat model.

Network Centric Warfare Modeling

A newer set of models has been described by researchers attempting to add more specificity to the concept of Network Centric Warfare (NCW) , the predominant theory of warfare in the Information Age introduced by Cebrowski, Garstka and Alberts 1998. They describe how the military must shift from platform-centric to network-centric combat, drawing a parallel in warfare to the use of information technology in the business sector (a process of shifting from platform-centric computing to network-centric computing). They describe the power of network-centric warfare as being governed by Metcalfe's Law, so that the "power" of a network is related to the square of the number of nodes in a network. This power comes from "information-intensive interactions" between the nodes. Cebrowski and Gartska describe how NCW results in an increase in speed of command, self-synchronization of forces, and higher situational awareness.[25]

Early attempts to model NCW used metaphors and thumb rules taken from the IT industry or attempted to re-cast traditional models as NCW models.[26] In general, the NCW literature has never graduated beyond loose metaphor or the type of "glittering generalities" that motivated frustrated Victorians to develop attrition-based models that are now dated.[27] In no case have the mechanisms of advantage for NCW been articulated with enough specificity to generate meaningful research, scientifically valid experimentation or rigorous concept development.

Some NCW modeling efforts to date include:

Use of IT industry models. The most prevalent of these is in the basic NCW text, which suggests that warfare will be conducted according to "Metcalfe's Law."[28] Recent research into network theory shows that this is a naïve assumption—networked behavior is far more complex then a simple count of potential connections.

Textual Descriptions. There are many cases in the Network Centric Warfare literature of imprecise textual models that do not hold up against more formal mathematic treatment. In an example from the basic NCW text, attempts to describe self-synchronization in detail assert that rule sets and shared awareness produce self-synchronization. Counter to this assertion, however, is the research that mathematically proves self-synchronization can occur without a common rule set or without shared awareness.[29]

Booz-Allen & Hamilton's Entropy Based Warfare Model. ™ At the core of this simulation are Lanchester's attrition-based equations with additional tuning parameters. Ironically, if one knew the value of the tuning parameters there would be no need for the attrition model. This model is also a poor representation of combat entropy.[30]

RAND studies on NCW measures of effectiveness (MOEs) for the Army and Navy. These efforts suffer from the same deficiency as the Entropy

Based Warfare model—they attach Information Age tuning parameters to what is essentially an Industrial Age model.[31]

Description of "Netwar" by Arquilla and Ronfeldt.[32] Although this work is valuable for its use of networks as metaphors, its textual descriptions of the dynamic behavior of networks do not correspond with the technical literature.

Research on Complexity Theory and Network Centric Warfare by Moffat.[33] This work disregards whole thematic topics in complex systems research and focuses too narrowly on agent-based modeling and the RAND NCW research mentioned above.

As these examples show, the NCW modeling efforts to date are also unsuitable candidates for an Information Age combat model.

TRANSFORMATION IN MODELING PHILOSOPHY

The basic structure of contemporary combat models is over 100 years old and a direct product of Victorian Era science, technology and philosophy. These models, reviewed in the previous sections, are inadequate for representing Information Age warfare for three main reasons:

(1) The models rely on mathematics that represent combat activities as independent processes. Networked processes are by definition interdependent.

(2) The models aggregate and dis-aggregate in a way that treats fine-scale behaviors as noise at the aggregate level. Such a process cannot adequately represent local tactical arrangements, clever use of information or massed effects from distributed forces. These, of course, are each important Information Age Warfare precepts.

(3) The models do not reflect the fact that the distribution of networked performance is highly skewed. Network-enabled feedback and feed-forward mechanisms can create increasing returns—"tipping point" behaviors—within Information Age systems.[34] Although NCW concepts are said to capitalize on these types of networked effects, contemporary model assumptions deliberately inhibit them.[35]

An Information Age combat model will require a transformation in military modeling philosophy and must address these challenges by explicitly representing interdependencies, properly representing complex local behaviors and capturing the skewed distribution of networked performance. In addition, an Information Age combat model must capture the dual combat processes important to distributed, networked warfare: the processes of attrition as well as search and detection. Such a model would be a bona fide transformation in combat modeling and constitute a true Information Age Combat Model.

[1] CEP is defined by the radius of a circle around an aimpoint in which 50 percent of a weapons system's rounds are expected to impact.
[2] See http://www.defenselink.mil/transcripts/2003/t03202003_t0319/effects.html (accessed 03 Jan 2005) for a recent transcript and slide briefing on EBO.
[3] Scott E. Page, "Uncertainty, Difficulty and Complexity," Santa Fe Institute Working Paper 98-08-076, 12 June 1998.
[4] Claude E. Shannon, "A Mathematical Theory of Communication," The Bell System Technical Journal, Vol. 27, pp.379-423, 623-656. See http://cm.bell-labs.com/cm/ms/what/shannonday/paper.html (accessed 05 Jan 2005).
[5] Technically, however, if there is any randomness in the process, an observer would have to read an infinitely long string of sentences to prove that they were created with a particular rule set.
[6] http://www.nist.gov/dads/HTML/zipfslaw.html (accessed 05 Jan 2005) discusses Zipf's Law.
[7] Data courtesy of Joe Bolmarchic, Quantics, Inc., from his briefing, "Who Shoots How Many," WINFORMS Conference, 22 Jan 2003 and "Who Shoots How Many," Proceedings, MORIMOC II, Military Operations Research Society, 1988.
[8] Substantial portions of this chapter (the next eight sections) are excerpted from Jeffrey R. Cares, An Information Age Combat Model, unpublished US DoD White Paper, 30 September 2004.
[9] See https://www.dmso.mil/public/transition/vva/ (accessed 01 Oct 2004) for a review of the US DoD VV&A program. In practice, funding for VV&A is almost never included in defense analysis contracts.
[10] Attributed to the statistician George E.P. Box. See http://en.wikiquote.org, accessed 01 Oct 2004. Thanks to Major Alistair Dickie, Royal Australian Army, for chasing down the source of the Operations Research "street wisdom."
[11] Some models attempt to represent as much detail as computationally possible. These "virtual models" are more useful as simulators than

simulations, that is, they are more useful for generating simulated experience (such as for pilot training) than for analysis.

[12] Shpak, M., Stadler, P. F., Gunter, P. W., and Hermisson, J, Aggregation of Variables and System Decomposition, Santa Fe Institute Working Paper 2003-04-25, April 2003, http://www.santafe.edu/research/publications/wplist/2003, accessed 30 Sep 2004.

[13] See for example, James J. Schneider, The Exponential Decay of Armies in Battle, Theoretical Paper No. 1, U.S. Army School of Advanced Military Studies, 1985.

[14] Liebholz, S. W., "Twenty Questions," in Wayne P. Hughes, Jr., (ed.), Military Modeling, Arlington, VA, Military Operations Research Society, 1984, 344-345.

[15] That is, the Central Limit Theorem applies. See http://mathworld.wolfram.com/CentralLimitTheorem.html, accessed 30 Sep 2004.

[16] These equations were first derived by a US Navy Lieutenant, J.V. Chase, in 1902. They remain attributed to Lanchester because Chase's work was classified until 1972. See Bradley A. Fiske, The Navy as a Fighting Machine (rev. ed.), (Annapolis: United States Naval Institute, 1988), 375-376. The equations were also derived independently by the Russian mathematician Osipov in 1913.

[17] See Bracken, J., Kress, M, and Rosenthal, R. E. (eds.), Warfare Modeling, (Danvers, MA, Wiley and Sons, 1995) for a listing and discussion of Lanchester variants.

[18] See https://www.jointmodels.mil/index.cfm?id=TACWAR/index.cfm for a description of TACWAR, accessed 01 Oct 2004.

[19] See http://www.metsci.com for a description of NSS, accessed 01 Oct 2004.

[20] See http://www.eadsim.com/EADSIMBrochure.html for a description of EADSIM, accessed 01 Oct 2004.

[21] See http://www.msiac.dmso.mil/spug_documents/JWARS_Overview_Brief.ppt for an overview of JWARS, accessed 01 Oct 2004.

[22] Wayne P. Hughes, "A Salvo Model of Warships in Missile Combat Used to Evaluate Their Staying Power," *Warfare Modeling,* (Danvers, MA: John Wiley & Sons, Inc., 1995), 121-143.
[23] Michael Johns, "Heterogenous Salvo Model for the Navy After Next, Master's Thesis," Operations Research Department, Naval Postgraduate School, 2000.
[24] Cares, Jeffrey R., "The Fundamentals of Salvo Warfare, Master's Thesis," Operations Research Department, Naval Postgraduate School, 1990 and Kieth J. Ho, Captain, Singapore Army, "An Analysis of Distributed Combat Systems," Master's Thesis in Systems Integration, 2001.
[25] Arthur K. Cebrowski, and John J. Garstka, "Network-Centric Warfare: Its Origin and Future," (U.S. Naval Institute Proceedings, January 1998).
[26] David S. Alberts, John J. Garstka, and Frederick P. Stein, *Network-Centric Warfare: Developing and Leveraging Information Superiority,* (Washington, DC: National Defense University Press, 1999). See also 2d ed. Rev., 2001. Available online: http://www.dod ccrp.org/NCW/ NCW_report/start.htm; and David S. Alberts, John J. Garstka, Richard E. Hayes, and David A. Signori, *Understanding Information Age Warfare,* (Washington, DC: CCRP Publication series, 2001). Available online at http://www.dodccrp.org/NCW/ NCW_report/ start.htm. All sites accessed 01 Oct 04.
[27] LT Chase created his equations out of frustration at the hand-waving and imprecise language that often accompanied discussions of the virtues of massed fires. The modern reader will note that Chase's disdain for "glittering generalities" would seem appropriate for NCW discussions today.
[28] Alberts, et al, 250-256.
[29] James Crutchfield and Yuzuru Sato, Coupled Replicator Equations for the Dynamics of Learning in Multiagent Systems, SFI Working Paper 02-04-017, 2002 and Robert Axtell, Non-Cooperative Dynam-

ics of Multi-agent Teams, Brookings, 2002, are two counterexamples to NCW self-synchronization claims.

[30] See Jeffrey R. Cares, "The Fundamentals of Salvo Warfare," Operations Research Department, Naval Postgraduate School, 1990, for a treatment of combat entropy that depends on the arrangements of assets, not Information Age attrition. See www.dtic.mil/ doctrine/jel/ jfq_pubs/1620.pdf, accessed 30 Sep 2004, for a description of the Entropy Based Warfare Model.

[31] Richard Darilek, Walter Perry, Jerome Bracken, John Gordon, and Brian Nichiporouk, *Measures of Effectiveness for the Information-Age Army,* (Santa Monica, CA: RAND, 2001); and Walter Perry, Robert W. Button, Jerome Bracken, Thomas Sullivan, and Jonathan Mitchell, *Measures of Effectiveness for the Information-Age Navy: The Effects of Network-Centric Operations on Combat Outcomes*, (Santa Monica, CA: RAND, 2002).

[32] David Ronfeldt and John Arquilla (Ed's.), *Networks and Netwars: The Future of Terror, Crime, and Militancy,* (Santa Monica, CA: RAND, 2001).

[33] Moffat, James, *Complexity Theory and Network Centric Warfare*, (DoD Command and Control Research Project, National Defense University, Washington, DC, 2003)

[34] Indeed, it has been long known that combat performance is better represented by skewed, rather than regular, distributions. See Joseph Bolmarchic, "Who Shoots How Many," Proceedings, MORIMOC II, Military Operations Research Society, 1988.

[35] As mentioned above, standard stochastic simulations are built to require only a relatively small number of runs for statistical convergence. A simulation that models skewed performance would require an exhaustively large number of runs.

3

Complex Military Environments

Many concepts for future warfare hope to extrapolate from the successful use of IT in air defense to other types of warfare by "digitizing" the battlespace. These approaches to automated combat assume that ubiquitous sensing linked to common operational pictures, centralized data bases and computers that can automatically assign weapons to targets will decrease reaction time, increase accuracy and optimize the weapons assignment process. Implicit in such assumptions is the belief that command and control is largely a matter of technology. A closer look at two different warfare processes, however, shows how military environments—the combination of natural context, man-made objects and competitive conditions in the battlespace—can have as much to do with command and control as technology by itself.

Air defense operations consist of high-power radio-frequency (RF) signals bombarding metal objects in a thin atmosphere. While air defense can certainly get complicated, the environmental conditions in this battlespace are nonetheless simple. A sure test for simplicity is to examine the operational representations at all levels of command. On a missile-defense ship, for example, one of the most junior watchstanders is a sailor who maintains the digital identification of radar returns. The most senior watchstander is the Tactical Action Officer (TAO), responsible for directing the ship's weapons. More senior to the ship's TAO is the TAO for the entire battle group, responsible for directing the sensors and weapons of all the air defense units in the group. Overseeing air defense operations is a one- or two star admiral. The digital combat displays monitored by all these watchstanders are virtually identical, with minor differences stemming from personal display preferences. This test shows that air defense is conducted in a *scale invariant* environment: scale invariant environments appear identical at different scales of observation.

Contrast this with a wholly different warfare area, Undersea Warfare (USW). In this environment, the most junior watchstanders, such as a Second Class Petty Officer sonar operator or maritime patrol aircraft crewman, each see different pictures. The former may be examining un-processed return on a sonar scope and the latter may be focused on sonobuoy contact information. Not only do these watchstanders each see different information, but a USW ship's TAO sees a different, ship-centric tactical view that includes local information like the sonic signature of neutral ships in the vicinity or the effects of *in situ* hydrographics. The admiral's staff, by contrast, might be monitoring fused information on a manually updated large scale chart while the admiral may only require a symbol on a digital display that indicates the last known time and position of a submarine detection. This is an example of a *scale free* environment: scale free environments appear different at different scales of observation. The best scales of observations in a scale free environment correspond to the scales at which decisions must be made. In a scale-free USW environment, fine scale information at the watchstander level is best for watchstander decisions. Coarser scale observations best support coarse scale decisions by more senior personnel.

The best method of command and control, then, can be determined not so much by technological possibilities as by the information conditions in the environment. Air defense experts will agree that their environment places a premium on reaction rather than adaptation. They operate in scale invariant environments where traditional approaches to risk and uncertainty dominate. Distributed networked operations are intended for scale free environments where complexity and adaptation are required. This chapter will define and discuss the characteristics of complex environments as a first step toward addressing adaptive command and control.

COMPLEXITY AND SCALE

The complementary concepts of complexity and scale help formalize Information Age command and control discussions. One way to measure the complexity of a system is to measure the number of ways a system can be usefully described. Scales are the levels of resolution at which descriptions of a system can be found.[1] Figure 3.1 describes the relationship between complexity and scale.[2] The number of possible interactions in a system is plotted on the horizontal axis. The number of ways the system can be described is on the vertical axis, ranging from low (simple) to high (complex). Consider a well-ordered system, such as a group of soldiers following very simple rules: they are each aligned behind the person in front of them, stay abreast of the person beside them and march in time with cadence. A single soldier, the guidon, serves as a reference point and one of the soldiers calls the cadence. A description of this system can be made quite concise: they are 100 soldiers marching in step. The individual behavior of soldiers need not be explicit for us have a good model of this system, so the number of ways the system is usefully defined is quite low. The system is therefore a simple system.

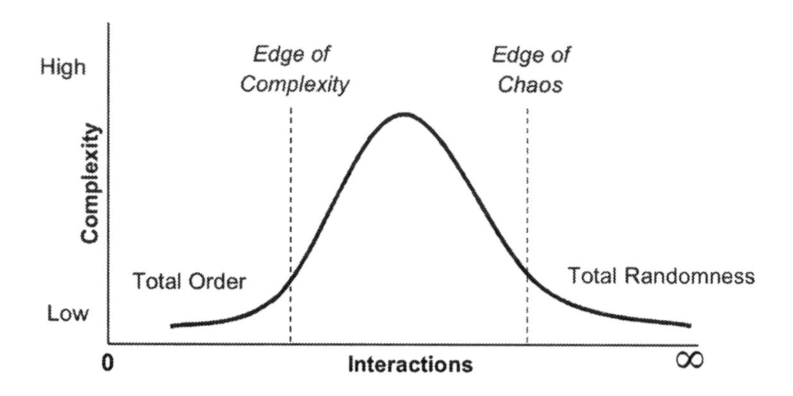

Figure 3.1—Complexity and Scale

Consider also a poorly ordered system such as a group of soldiers aligned behind and staying abreast of whomever they choose, each acting as their own guidon and each calling their own cadence. Counter-intuitively, this system is just about as simple as the well-ordered system. A description of this system can also be made quite concise: they are 100 soldiers, none of whom are marching in step. It is not useful to describe the random march of all the soldiers in detail, so a simple, coarse scale representation will do. These two systems are found in the lower left and lower right tails of the hump-shaped curve in Figure 3.1. A more complex system could be comprised of, for example, 10 mutually supporting squads of 10 interdependent soldiers executing squad tactics. This longer system description would be found closer to the peak of the curve in Figure 3.1. There are many more ways this complex system can be described and no single scale of observation can be a complete description. Coarse scale observation might highlight the overall pattern of inter-squad cooperation while medium scale observation might bring the cooperative efforts of two of the ten squads in to focus as well as some of the specific interactions between these two squads. The tactics within each squad can be observed at an even finer scale.

MULTI-SCALE REPRESENTATIONS

To illustrate these ideas with a more realistic military example, consider a very large amphibious landing. At the three-star general's level, the landing can be described in terms of ships, objectives, Marine Expeditionary Units (MEUs) and aircraft sorties available. From the perspective of an amphibious ship captain or MEU commanding officer there is much more detail, including transit lanes, synchronized waves, landing zones and flight schedules. In addition, there are some issues and decisions at the three-star scale that are not relevant, useful or perhaps not even observable at the commanding officers' scale. From the individual Marines' viewpoints, there is a host of detail that changes rapidly from one moment to the next, particularly while they cross a hostile beach under fire or egress a transport helicopter in a contested landing zone. This fine scale detail, while important to the individual Marines, is not useful at the scales of observation of a commanding officer or of the three-star's staff.

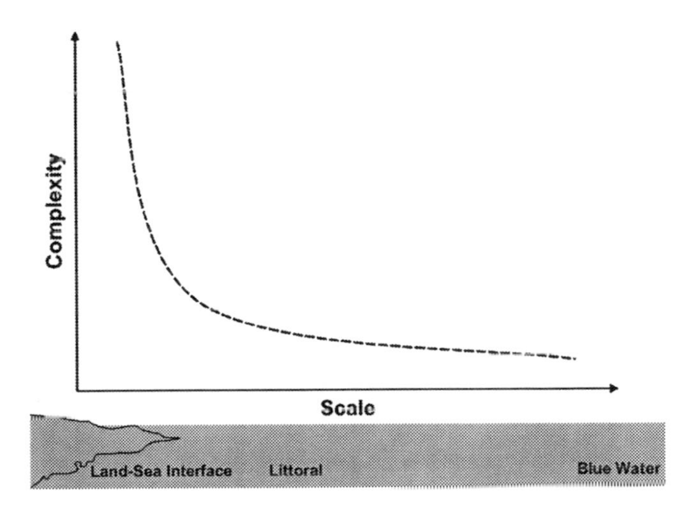

Figure 3.2—Multi-scale Representation of a Littoral Battlespace

Figure 3.2 portrays the relationship of complexity to scale in littoral operations such as this notional amphibious operation. It has a curve typical of systems with a great deal of complexity in one part, moderate complexity in another part and relative simplicity in yet another part.[4] The curve is "scale free": there is no single scale at which the all the important system behaviors are described. Systems with such complexity-scale profiles are called *scale free systems.* These systems require *multi-scale representation* to show behaviors and structures at the scale at which they are most informatively displayed.[5] Another way to think of this curve is as a description of all the information required by all decision makers at each level in a complex military organization.

"No Free Lunch" Theorem For Control

In mathematics and computer science, optimization is the analytical process of finding solutions to problems in the shortest time possible. Researchers have recently derived the "No Free Lunch" (NFL) Theorem that proves there are no universally best optimization routines. Algorithms perform well on one type of problem at the expense of poor performance on other types of problems because the success of an optimization routine depends on the extent to which the structure of a routine matches the structure of a problem.[6] In other words, the scale of the best routine is the same as the scale of the problem. Therefore, there exists no universally efficient problem-solving routine for ensembles of problems that are scale free, only routines that work best on different parts of a collective. A collection of different routines that work best on different parts of a collective problem would, however, exhibit the scale free property.

Identical logic can be applied to command and control. If an operation is scale free, then local behaviors occur at scales and complexities substantially different from global behaviors. An optimized control process cannot, therefore, control all local and global behaviors. By treating adaptive operations as a "problem" to be solved by a command and control method, one can propose a similar No Free Lunch Theorem for Control: the best control method is one in which the complexity of the control processes most closely matches the complexity of the operations.[7]

Associating the degree of an operation's complexity to the ease by which it can be controlled by a regulator has a pedigree in Control Theory, where it is discussed as the Law of Requisite Variety.[8] The Law of Requisite Variety states that a controller must be able to cope with at least as much complexity as the system it controls.[9] This supports the notion that the best control mechanism is one that matches the complexity of the controlled operation.

If an operational force exists merely to train in peacetime garrison, then one might call this a "closed system," meaning that exogenous factors do not greatly impact an observer's ability to monitor the operation, decide how to control the operation and then successfully control it. Combat operations (as well as non-combat operations such as peacekeeping and humanitarian assistance) are "open systems"; adversaries and the physical environment can thwart even the best-designed plans.[10] Open systems are typically more complex than closed systems; most combat operations will therefore require complex control and multiscale observation.

Domains of Warfare

Military competition can unfold in many dimensions. One way to improve the study of these dimensions is to group different environments in different *domains*. Think about how an observer perceives a collection of physical objects. The collection itself is operating by a set of internal rules, the *causality* within the collection. Causality cannot be directly observed, however, but neither can the physical objects. Observers never actually observe physical objects, they observe phenomena emanating from physical objects, such as visible light bouncing off surfaces, heat detected by infra-red sensors, communications signals, RF energy reflected off objects by a radar system, etc. From these phenomena, the observer must develop a set of propositions not only about the physical constitution of the physical objects, but also the causality itself. In effect, the observer is *encoding* signals from the environment into a description of the collective.

These propositions might include such elements as the presence of tracked vehicles, the fact that the tracked vehicles are all moving west at a high rate of speed or that there are slow moving aircraft in the vicinity. Collectively, these propositions become what Information Theory calls a *formal axiomatic system.* If the propositions in the formal axiomatic system begin to converge to a coherent, plausible description of the behavior of the physical objects, then the observer might claim to have derived the *implication* of the signals. The goal of command and control is to achieve close correspondence between the implication of signals encoded from a physical system and causality motivating the physical systems. In a military example, a formal axiomatic system would be a collection of intelligence reports describing a set of physical objects, implication would be an intelligence officer's assessment of the likely intent or rule sets behind the physical objects' behavior and causality would be the actual intent or rule sets (known, of course, only within the observed collective but not to the observer). If the observer also seeks to act, then the observer must use the implication to *decode* from the formal axiomatic system into a signal or action that can influence the physical objects in terms that the physical objects can perceive. Figure 3.3 portrays how these processes interact in what the literature of adaptation calls the Modeling Relation.[11]

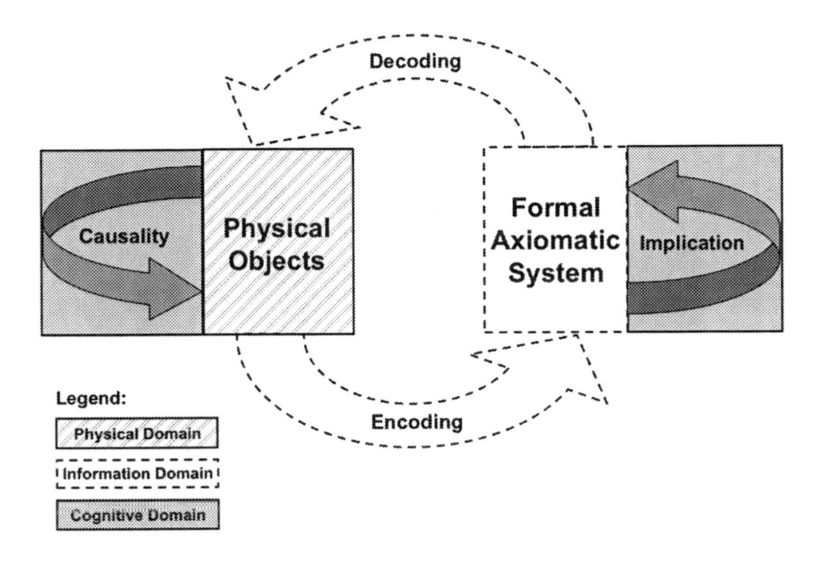

Figure 3.3—The Modeling Relation

The elements of the Modeling Relation are each measured and understood with their own unique scientific and mathematical approaches, providing the following analytical parsing. The *Physical Domain* consists of all tangible elements in an environment (importantly, this includes IT structures). The *Information Domain* encompasses the encoding and decoding processes, as well as the formal axiomatic system. The *Cognitive Domain* consists of the implication (human and automated interpretations of the Information Domain) and the causality within the observed system. Note that the Figure 3.3 portrays only one side in a competitive process. This figure could be modified to include multisided competition.

Figure 3.3 contrasts with the way that domains of warfare have been defined elsewhere in the Information Age military literature. Other definitions suffer the same fate as early attempts at developing the science of Biology: they are the result of categorization, not analysis. For example, one construct in the literature places IT hardware in an information domain although it is clearly tangible, and observations bypass an information domain altogether to feed directly into "human perception" (which is not contained in any domain at all).[12] The parsing contained in Figure 3.3 predates this literature[13] and derives the groupings using the scientific literature of such topics as semiotics, Information Theory and complex systems research rather than arbitrary categorizations.

A few comments about the domains are appropriate. First, because of advances in IT and robotics, elements in an Information Age Physical Domain can be very widely distributed in space, throughout time and within an organization. This makes derivation of the implication much more difficult than in the Industrial Age. Massed forces headed at high speed toward an international border, for example, are not ambiguous, but staggered movement of an opponent's small units to dispersed locations certainly is.

Second, although warfare is changing in all three domains, the Physical Domain attracts more of the innovator's attention. A main reason for this preoccupation with the Physical Domain is that our Industrial Age education and professional culture reinforce physical models of systems. Most important mechanisms of advantage for Industrial Age warfare have been found in the Physical Domain and a characteristic of Industrial Age systems is that the information and cognitive processes typically correspond closely to the physical processes in a system. We have therefore been very successful using physical models as proxies for information and cognitive models, but this will not likely be true in the Information Age. Military operations are complex precisely because of interactions and interdependencies in the information and cognitive domains. Although most innovators are focusing on physical platforms and physical IT systems, true transformation is the result of innovation in the information and cognitive domains.

Third, although highly specialized (yet incomplete) theories of information exist to describe functions in the information domain (such as

Shannon Information Theory and Algorithmic Information Theory), no "science of information" currently exists. Some of the most advanced research institutions in the world have set out to correct this scientific deficiency. Multi-disciplinary teams have begun to discover and exploit deep relationships between quantum mechanics, thermodynamics and information theory that will have a significant impact on the ability of systems to fuse multi-spectral sensor data, resolve complicated schemes for distributed computation and provide a new generation of cryptology.[14]

Fourth, while the choice of the term "formal axiomatic system" might seem like an unnecessarily awkward use of jargon, it is useful because it invokes an important theoretical concept. In 1931 the Austrian mathematician Kurt Gödel proved that any formal system of axioms is incomplete, that is, one can never be sure that introducing a new axiom into the system won't create inconsistencies within existing axioms and thereby invalidate the system itself.[15] This, of course, has direct military application to deception, surprise and information fusion. While many advanced military systems call for a process that looks very much like the Modeling Relation, none approach this type of incompleteness with appropriate humility.

Finally, the study of the human mind (and collectives of human minds) is one area for which science provides few answers. The inner workings of the human brain are still largely a black box to researchers. This has been identified as a new frontier in Information Age research.[16] Answers to questions about distributed commander's intent, shared awareness, speed of command and self-synchronization depend on advances in the cognitive sciences. Applying cognition to a dynamic process is the essence of control. The next chapter will address the fundamentals of adaptive control in complex environments.

[1] For a full mathematical treatment of complexity and scale in complex systems see Section 8.3 of Yaneer Bar-Yam, *Dynamics of Complex Systems*, (Addison-Wesley, Reading, MA, 1997).
[2] W.H. Zurek, "Algorithmic Information Content, Church-Turing Thesis, Physical Entropy, and Maxwell's Demon," Zurek, ed., *Complexity, Entropy and the Physics of Information*, (Addison-Wesley Publishing Company, New York, 1991).
[3] The next two sections originally appeared in Jeffrey R. Cares and CAPT Linda Lewandowski, USN, Sense and Respond Logistics: The Logic of Demand Networks, undated, unpublished US Government white paper, 2002, for a discussion of this concept.
[4] For a complete discussion of the impact of complexity and scale on littoral operations, see Yaneer Bar-Yam, "Multiscale Analysis of Littoral Warfare," CNO Strategic Studies Technical Paper, 2002.
[5] The concept of multi-scale representation has direct application to military command and control problems. For a short definition of this concept, see http://www.necsi.org/guide/concepts/multi scale.html, accessed 30 March 2004.
[6] David H. Wolpert and William G. Macready, "No Free Lunches For Search," Santa Fe Institute Working Paper 95-02-010, http://www.defenselink.mil/transcripts/2003/t03202003_t0 319/effects.html, accessed 30 March 2004.
[7] Cares, Jeffrey R. and CAPT Linda Lewandowski, USN, Sense and Respond Logistics: The Logic of Demand Networks, undated, unpublished US Government white paper, 2002.
[8] W.Ross Ashby, Introduction to *Cybernetics,* Part II, Variety, (Chapman & Hall, Ltd., New York, 1957).
[9] http://artsci-ccwin.concordia.ca/edtech/ETEC606/principles/law.html, accessed 30 March 2004 has a discussion of the Law of Requisite Variety and some useful links.
[10] The case for treating military systems as open systems is made by the Deputy Director, J8 (Wargaming, Simulation & Analysis), The Joint Staff, Vincent P. Roske, Jr., in "Opening Up Military Analysis: Explor-

ing Beyond The Boundaries," Phalanx, (Online) June 2002, Volume 35, Number 2, http://www.mors.org/publications/phalanx/ jun02/ lead.htm, accessed 30 March 2004.

[11] Robert Rosen, *Anticipatory Systems: Philosophical, Mathematical and Methodological Foundations* (IFSR International Series on Systems Science & Engineering, Vol 1) (Oxford, Pergamon Press, 1985), 45-220.

[12] David S. Alberts, John J. Gartska, Richard E. Hayes, and David A. Signori, *Understanding Information Age Warfare,* (National Defense University, Washington DC, 2001), 9-29.

[13] Jeffrey R. Cares and John Q. Dickmann, "Information Age Warfare Sciences," presented at Preserving National Security in a Complex World Conference, The Ernst & Young Center for Business Innovation, Cambridge, MA, 12-14 September 1999.

[14] "Making Complexity Simple," http://www-mtl.mit.edu/%7Epenfield/pubs/complex-99.html accessed 26 December, 2000.

[15] See http://mathworld.wolfram.com/GoedelsIncompletenessTheorem.html accessed 30 December 2004.

[16] See http://www.gmu.edu/departments/krasnow/, accessed 26 December, 2000.

4

Adaptive Command and Control

This chapter on adaptive command and control begins with a discussion of the opposite of adaptation, optimization. Indeed, pure adaptation and pure optimization cannot coexist in a system. In the domain of practical management, optimization is closely allied with control—managers seek to optimize a system mainly to influence or control the things they manage. Indeed, when a system is deemed "optimized," modern managers assume events and processes are in a sweet spot of productivity, efficiency and profitability. An optimal system implies that "things are under control." If a system becomes less than optimal, managers attempt to regain control by identifying and exploiting sources of even greater efficiency.

Optimization, however, can fail as a method of control in uncertain, dynamic environments. Two facts provide evidence for this assertion. First, strict optimization is best applied to very stable systems, usually those that are highly engineered or are not subjected to complex environments (that is, "simple" systems). Second, the best control mechanism is one that matches the complexity of the controlled system. These two complementary ideas inform a new perspective for command and control of adaptive operations.

OPTIMIZATION V. ADAPTABILITY

The extent to which a system can be optimized depends on more than just the instabilities and complexities within the system itself. The capacity for a system to be optimized also depends on the instabilities and complexities within other systems to which the system is connected and instabilities and complexities within its external environment. Paradoxically, in very unstable or complex competitions and environments, too much efficiency can actually impair the ability of a system to function properly. To see how this might happen even in a very simple, linear system, consider the problem of *makespan* in production line fabrication.

A block of metal that proceeds through a sequence comprised of several milling steps that create, say, a widget, is the object of a process that literally "spans the making" of the widget. If a certain amount of time is required on each machine, then the total milling time plus transfer and set up time is the total makespan. For example, if the widget required ten one-minute milling steps and one minute to transfer and setup between milling machines, then the total makespan would be 19 minutes. Transfer time and setup time are generically called *slack time*. A process with sufficient slack time is tolerant of small failures like temporary machine malfunctions or operator errors. Sufficient slack time means that operators can work around these failures (by fixing malfunctions or correcting the errors) and put the process back on schedule. Too much slack time, however, is wasted overhead. Managers of production lines create optimally controlled processes by removing as much unnecessary slack time as possible.

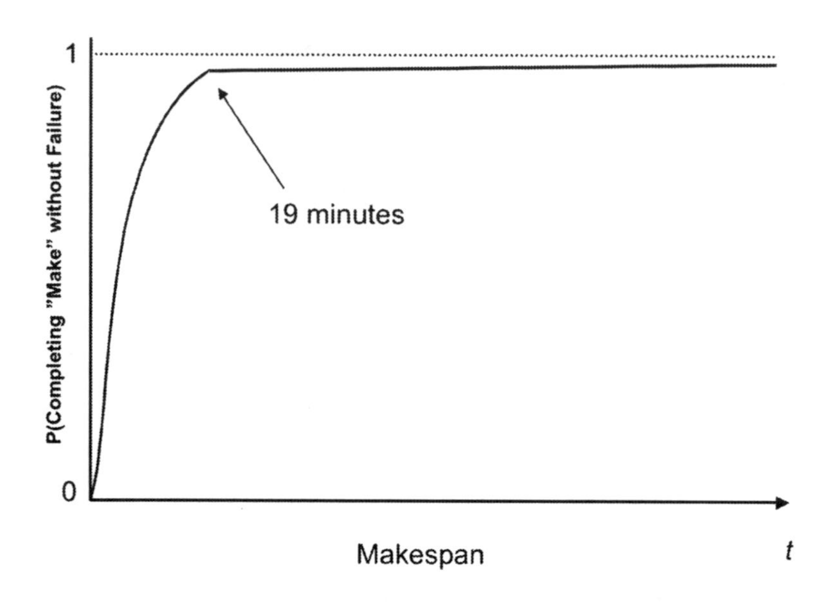

Figure 4.1—Makespan and P(Process Failure)

In this example, as slack time approaches nine minutes, the chances that the entire process can fail starts to increase. Makespan under 19 minutes dramatically increases the potential for failure. Figure 4.1 graphs a curve of the relationship between makespan (horizontal axis) and the probability that a process is completed without failure (vertical axis).

The analogy to command and control is direct: if too much slack is removed from the command and control process, military units become much less tolerant of change and missions can fail catastrophically. Of course, adaptive operations are much more complex than the finely engineered, simple process of production line fabrication. Information Age warfighting concepts suggest that fluid, self-synchronizing military forces will be the norm, at least at the tactical level. It has been proposed that the primary source of advantage in these self-synchronizing forces arises from networked effects that can be summoned for use in the manner of advantage chosen by clever commanders based on evolving conditions.[2] Optimized command and control methods with too little slack will poorly serve attempts at self-synchronization in complex environments. The makespan analogy applied to distributed networked operations justifies the first main point, that strict optimization is best applied to very stable systems, usually those that are highly engineered or are not subjected to complex environments.

Complex Control

The need for complex control of adaptive military systems has never been greater. Although rudimentary military networks have existed for many decades, preliminary concepts for advanced networked forces began to emerge in the early 1990s. These concepts mirrored a similar phase of technological innovation in other industries: the initial thrust was in developing systems to help people perform current tasks more efficiently with IT. One drawback to this approach is that there is always an upper limit on efficiency—merely investing in IT without substantially changing how current tasks are performed can result in the removal of too much slack from a system. Early network centric warfare concepts were all innovations of this type.

The defense community has recently started to re-think its focus on merely enhancing military systems with IT. New concepts for smaller distributed forces are beginning to emerge. The basic assumption behind these concepts is that distributing a military force creates more options for a commander, increases the surveillance burden of an adversary and allows for massed fires while forces remain dispersed—all while requiring less force protection. Without proper networking, however, distributed forces are vulnerable to destruction or mission failure. There is, therefore, great advantage to both networking and distributing. Methods of sense-making and control in complex systems will therefore continue to grow in importance as concepts for future distributed networked forces mature.

The greatest obstacle facing these new concepts is that the military is still a strong adherent of control by optimization. Command and control of distributed networked operations will require new notions for adaptive control of distributed collectives in complex environments. The remaining sections of this chapter present a new concept for adaptive command and control. The sections relate complex control theory to the use of rule sets and contain four main topics. First, a general theory of complex control is developed and discussed. Second, the mechanisms by which systems create simple rule sets for responsive adaptation are examined. Next is a discussion of the cycle of alternately gaining and losing control that systems inevitably experience when they interact with the real world. The chapter concludes by presenting the results of recent research into types and uses of rules.

A General Theory of Complex control

One of the biggest challenges of command and control in distributed networked operations arises from the fact that not only are behaviors occurring at many different scales, but that many of the behaviors are the result of actions that one side itself cannot control: the actions of the adversary.[3] Controlling such a system requires an ability to discern patterns in the observed dynamics of the system (including both signal and noise). Some types of signals are usually evident and known to both sides. These signals include information about the basic structure of the environment and nature of the competition, as well as some base level of information about the competitors themselves. This is information that is not likely to change during the time scale of the competition.

After basic patterns are determined, all other observations might appear random (noise). However, within this "apparent noise" are two other types of information: dynamic patterns yet to be discovered and true noise.[4] As more observations are made, the goal is to extract signal out of the apparent noise and combine it with the basic structural information to create an even clearer picture of the competitive landscape. This "learning through feedback" is central to success in controlling complex systems and is one of the hallmarks of distributed networked operations.[5] To achieve a deeper and more formal definition of adaptive control in complex systems requires non-traditional ways of describing pattern, signal and noise. The following presents the underlying technical arguments of complex control and begins with some basic definitions:

Entropy is variously defined as a measure of wasted effort, lost energy, uncertainty or randomness.[6] In the context of this paper, entropy is an inverse measure of how much an observer understands by observing military competition in a complex environment. Entropy is said to be high when little is understood about the competition.

Shannon Entropy (H) can be defined as a measure of the acuity with which an observer receives a signal.[7]

Algorithmic Information Content (K) measures the extent to which a received signal can be compressed into a more compact description of the signal. It is sometimes defined as the shortest bit stream that can be used to describe a signal.[8]

Total System Entropy ($S = H + K$) is a measure of the amount of information gained by observation. Entropy is high when there are high levels of noise relative to the levels of signal or when there is no pattern in observed signals. A decrease in entropy is occurs with an increase in understanding.

Noise ($N = 1\text{-}S$) is randomness that contains no information about military competition in a complex environment.

Figures 4.2 and 4.3 show how these definitions are related. Consider a bit stream of finite length n that is observed over some time, t. As t progresses, more information is received by the observer about the bit stream. If there were a perfect transfer of bits to the observer (that is, a noiseless channel from the system to the observer), at some t all n bits would be received and Shannon Entropy, H, would be zero. Figure 4.2 shows the case when the entire bit stream has been received but the bit stream is completely random. In this case, although the Shannon Entropy is zero, the Algorithmic Information Content, K, is at a maximum of 1.0—there is no shorter way to describe a random bit stream than by using the entire bit stream. In other words, the Total System Entropy is preserved, indicating that there is no pattern for the observer to discern, even when the bit stream is observed with perfect clarity. Another way to think about this is that the observer does not have a description of the system that is simpler than the collective of all observations.

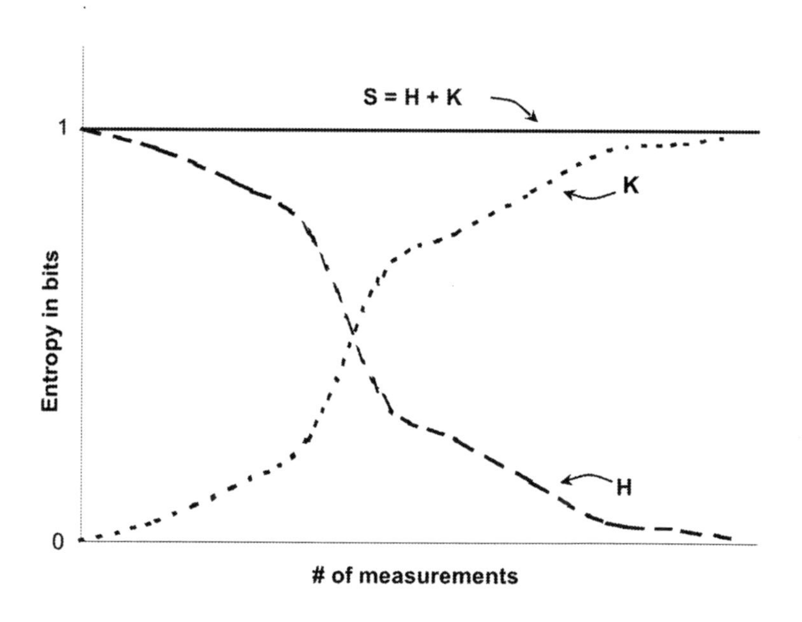

Figure 4.2—Total System Entropy

Figure 4.3 shows a more common set of observations. In this case, the Shannon Entropy is again reduced to zero but the Algorithmic Information Content is reduced as well—the observer discovers a more compact way of describing the bit stream. The Total System Entropy decreases, indicating that there is a pattern for the observer to discern and more is known about the bit stream after observation than prior to observation. Another way to think about this is that the observer now has a description of the system is simpler than the collective of all observations.[9]

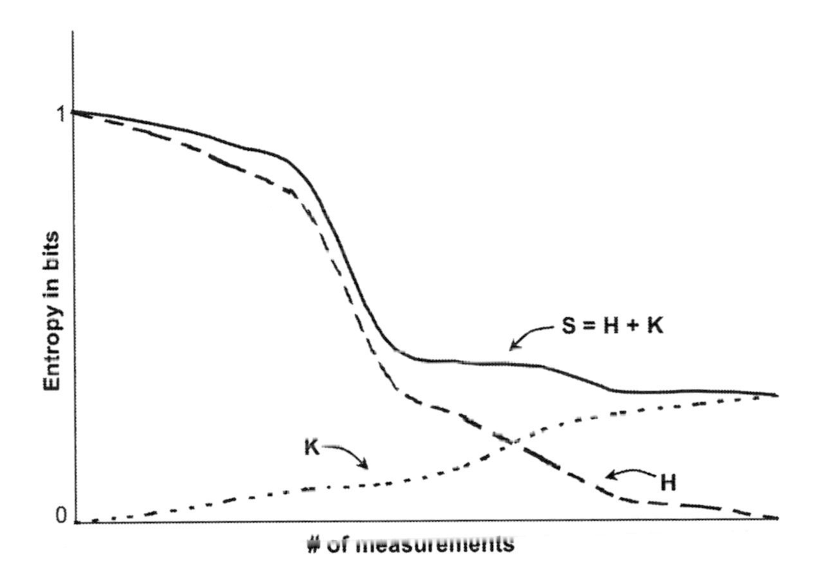

Figure 4.3—Total System Entropy

The systems described in these two figures have unique features that must be modified before the ideas can be more directly applied to complex control. First, the preceding discussion assumed that the bit stream was of fixed length. In military competition in complex environments, the amount of information is never fixed and may never be known by the observer. Indeed, deciding when sufficient information has been received can be one of the most difficult decisions in warfare. Second, the assumption in Figures 4.2 and 4.3 is that the bit stream is received over a noiseless channel. Noise, however, is a key feature in almost all competitive complex environments and discerning signal out of noise is another of the most demanding tasks in control of distributed networked systems.

H, *K* and noise interplay for complex control the following way. As an observer samples military competition in a complex environment, the observer attempts to combine signals into existing patterns. As a practical point, however, just how much of this originally assessed pattern is noise and how much is signal can never be known, because what appears as pattern early on might be rendered irrelevant (and therefore redefined as apparent noise) as additional signals arrive.[10] Control in complex systems is achieved through continuous observation, repeated comparison of new observations to existing patterns and direct application of new knowledge. As the competition or the environment become more complex, comparing signals and noise with existing patterns becomes increasingly difficult. The difficulty lies in fact that an observer must watch a system until infinity to completely specify which part of a system is pattern and which is pure noise.[11] No one can observe forever, so the complex control problem cannot be solved completely. The best one can do is to learn through feedback more quickly than the system and the environment changes. It may also be sufficient merely to learn at a faster rate than an opponent learns.[12]

It is therefore even more important to interact with a competitive system at as many scales as possible. This ensures that one observes important behaviors on the scale at which they occur. It also increases the chances of obscuring actions to a competitor who is not observing at a particular scale. Of course, controlling one's own system is much easier than trying to control an enemy's system, because even if plans are interrupted, friendly operations are still driven by commander's intent. The insight instilled in each decision maker by commander's intent can be a crucial advantage in complex competition, since in the midst of confusion, apparent noise is more likely to become signal when viewed through the lens of commander's intent. In other words, creating patterns from signal is easier if the observer is informed by *a priori* knowledge of a system's rule sets. By contrast, trying to control a system without foreknowledge of its causality is much more difficult.

Rule Sets: Internal Models for Adaptation[13]

The preceding discussion might appear to be no more than technical arcana but it is in fact a formal description of how humans cope during competition in complex environment. Humans translate observations of a competitive environment into an operative language with which to make decisions and direct action. This operative language is the language of rule sets.

Just what are these mechanisms for rule-making? The answer to this question lies in a closer examination of the relationship between a complex system and its competitive environment. No system of interest to the defense community exists in isolation—each system is contained in an environment, which except in the most trivial cases, includes competitors as well. As stated earlier, a system that adapts to its environment does so by learning. Some questions naturally follow:

- What does it mean to learn?
- How does a system learn?
- How can it be known if a system is learning at a rate that contributes to successful adaptation?

The preceding subsection partially answered these questions by suggesting that control of complex systems can be achieved by continuously observing and comparing new observations to existing patterns. This, however, is more of a goal statement than identification of a useful method for learning in a complex environment. As a first step toward a more useful method of learning, we must examine the relationship between what are known as the *external complexity* and the *internal complexity* of an adaptive system.

External complexity is a measure of "the input, information or energy obtained from the environment that the system is capable of handling or processing."[14] This type of complexity can be thought of as *data complexity*. Internal complexity is a measure of the system's ability to internally represent input, information or energy. This could also be called *model complexity*. Successful adaptation is achieved by increasing external complexity while decreasing internal complexity. In practical terms, this means observing and processing a great deal of information and converting it into simple rules. These rules can then be used to process an even greater amount of environmental and competitive information during subsequent observations.[15] This feedback loop is an example of a learning curve that explains how system performance can improve through experience.

Still, this is a description of what is to be done and not a method for how to do it. Key to arriving at a method is to recognize that external complexity and internal complexity are operative at different time scales. External signals and noise are observed individually until they can be identified as noise or associated with a pattern (either with a pattern derived by *a priori* analysis or patterns already formed through observation). Either way, there is a time lag between the receipt of a signal and its incorporation into an existing or emerging internal model of the external environment. This similar to the case in Figure 4.3, the model is a simpler description of the observations than the collective of all observations themselves.[16]

Learning, then, can be thought of as the "transformation of correlations into associations…and these associations serve to predict and anticipate future events."[17] In other words, learning is converting observations from a complex world into simple rule sets that help a system adapt. This is sufficient to answer the first question, "What does it mean to learn?" and contributes to answering the second question, "How does a system learn?" but does not yet constitute a specific method.

Interestingly, what might seem to be the hardest question, "How can it be known if a system is learning at a rate that contributes to successful adaptation?" provides insight into the seemingly easier question of how systems learn. Once again, examining the interplay between external complexity and internal complexity is instructive. If the external complexity of a system is lower than the internal complexity, the system has a complicated plan for a simple world. The system therefore learns less with each successive observation and eventually stops adapting.

If the external complexity and the internal complexity of the system are roughly equal, then the system has constructed a virtual model of the external environment. By making the internal view of the world no simpler than the signals received from the world, the system can adjust its performance and strategies no more quickly than the rate at which the world itself unfolds. This phenomenon supports one of the key concepts in this chapter: reactive response is not adaptive response and will likely provide insufficient control in complex environments. The best adaptation occurs when the external complexity remains higher than internal complexity, so that internal rule sets are robust even when signals from the external environment change at a high rate.

There are two important implications of this point, which are also evident in research into learning rates in complex systems. The first implication is that larger, more general rules are better than specific rules in complex environments. Over-complicated, nested rule sets create a virtual internal representation of the external environment and therefore only provide for adaptation at the same pace as change occurs in an environment, that is, they can only provide reactive response. Although general rules are less efficient, they are more robust to dramatic changes in system input or uncertain environments. Put differently, there is a choice between good-enough rules for most cases or perfect rules for cases that might not occur.

The second implication is that a small set of general rules that create robust behaviors most quickly are not usually refinements of each other. When more than one rule explains a significant amount of a system's total behavior, these rules must all be general rules. By definition, one general rule cannot be derived from another general rule. Otherwise, the rule that produced the derived rule would redundantly describe the same external complexity but with a higher internal complexity than the derived rule, so the producing rule would cease to be general. Using Set Theory, one would say that the most adaptive rule sets contain rules that are not subsets of each other.[18] In practice this means that after a general rule is determined to explain a portion of a system's behavior, efforts should be directed at determining another general rule to explain the remaining behavior rather than attempt to refine the first general rule.

A method for how a system learns and adapts can now be stated directly. An adaptive system takes specific, detailed information from its competitive environment and creates general rules. An adaptive system does not take general information and create specific rules, nor does it take specific information and create specific rules. Specific rules, as noted above, are a feature of reactive systems. Finally, an adaptive system does not take general information and create general rules, because that would mean the system operates in a simple environment for which adaptation is not required. A system successfully adapts by taking the complex and making it simple. In doing so, an adaptive system is in some situations less efficient than a specialized, reactive system, a fact that raises substantial cultural issues, since inefficiency is anathema to traditional managerial sensibilities that rely on optimization.

Adaptation Over Longer Time Scales

Any system of interest to the military is an "open" system interacting with a competitor in a complex environment. The Second Law of

Thermodynamics says that the information in an open system is diverging into many states. Effort expended for control of one part of an open system must always dissipate into the rest of the system. Radical changes to environmental conditions at all scales due to, say, innovation, surprise or blunder, can speed this dissipation. The Second Law therefore mandates that no rule set is forever robust. The environment and competitors can—and usually do—invalidate the best rules that humans or machines devise. When that occurs, the system must begin learning anew. This longer-term adaptation, the learning-relearning cycle, is also addressed in the complex control literature.[19]

Consider a system introduced into a complex environment with complete ignorance of the environment or competitors. Initially, this system would perceive all input as potentially relevant and internal complexity would be high (a pattern would not yet form) and without a good *a priori* model, the internal complexity would not quickly converge to a general rule set. Coarse scale exploration of the environment would be the most useful learning strategy because the system cannot yet process complex data (because it has no general rule set). Even when starting with coarse scale information about the environment or the competition, however, internal complexity can still decrease with subsequent observations.[20] An adaptive system can therefore emerge from complete ignorance and derive and refine general rules that are robust and ensure survival, so long as the time scale at which the environment changes is longer than the timescale at which the system learns. As the system learns through this initial coarse exploration, a longer time scale evolution from a state of low external complexity and high internal complexity to a state of high external complexity and low internal complexity can result.

But learning rates are not linear. An adaptive system can learn at a rate much faster than the rate at which both the environment and the competition change. If the environment or the competition changes much more slowly than an adaptive system learns, adaptation is not valued and specialization occurs. This is because a model of a stable environment can be created from simple observations. Since it need not be robust, this model can become specialized. But since this specialized model is not the product of complex inputs, the model can never become robust. If the environment suddenly becomes more complex the specialized model will eventually become invalid. When this occurs, the system must return to a general exploration of the environment to start the adaptation process all over again. This cycle from exploration to exploitation and specialization back to exploration is typical of complex control problems. Again, the need for general, robust *a priori* rule sets in complex environments is underscored.

This exploration-exploitation cycle has been observed in natural and simulated ant foraging. A typical cycle starts with the ant collective ignorant of food source locations. As ants leave the nest, they each begin a random search of the area around the nest. Initially, there is no pattern in the spatial location of the ants. Since a food source has not yet been found, the Shannon-type of entropy, which measures the amount of information the average ant has about food source location, is at a maximum. When an ant finds a food source, it follows a simple rule: take food to the nest and leave a pheromone trail en route. Other ants can sense the pheromone trail and from it can determine the direction of a food source. As their random walks lead them across the trail, the pheromone scent triggers another simple rule set: follow the trail to the food, take food to the nest and leave additional pheromone en route. Over time the pheromone trail gets stronger, so more ants find the food and return to the nest leaving more pheromone. The familiar pattern of ants shuttling from their nest to a food source emerges and the spatial entropy of the ant's collective behavior decreases. In addition, the amount of information about food source location that the average ant "knows" begins to grow, corresponding to a decrease in Shannon-type entropy about food source locations. When the collective has exploited all the food, however, ants no longer follow their exploitation rule set and pheromone starts to evaporate.

The ants' coherent pattern begins to break down and Shannon Entropy begins to increase. But the breakdown in pattern corresponds to a new random search that will locate additional food sources and eventually the exploration-exploitation cycle will begin all over again.

Ant foraging behaviors therefore are both produced by and create a natural cycle of adaptation over long time scales. A remarkable fact is that same rule sets that exploit the food source guarantee its depletion through non-linear positive feedback loops (an increasingly stronger pheromone trail as more ants follow it). Just as remarkable is the fact that the same rule sets that enable the ants to find the food help re-set the system once the food is depleted. Ant foraging is a good illustration of adaptive behavior and robust rule sets. It is worth noting that these rule sets were developed by perhaps millions of years of evolution, a process, of course, which entailed quite a bit of trial and error. Although very little research has been devoted to developing the processes to produce rule sets for adaptive command and control, some results do exist. The next section presents the results of recent research into a rule set generating system developed to support a DoD adaptive logistics concept.[21]

Types and Uses of Rules

Webster's New Universal Unabridged Dictionary defines the noun and verb "rule" variously as:

> "[noun] (1). An established guide or regulation for action, conduct, arrangement, etc. (2). A complete set or code of regulations… (3). A fixed principle that determines conduct; habit; custom… (4). A criterion or standard. (5). Something that normally or usually happens or obtains; the customary or ordinary course of events… (6). Government; reign; control… (8). Way of acting; behavior. (9). …a decision or order… (10). A method or procedure for computing or solving a problem.… Synonyms—government, sway, con-

> trol, direction, regulation, law, canon, precept, maxim, guide, order, method...
>
> [verb] (1). To have influence over... (2). To lessen or restrain.... (3). To have authority over; to govern; to direct... (4). to determine... (2). To stand at or maintain a certain level..." [22]

Clearly, "rule" is a word that enjoys rich usage in the English language.[23] All these meanings should be included in the language of distributed networked operations, yet there is a world of operational difference between, for example, a "regulation for action" and something invoked to "lessen and restrain." One sense of the word seems to suggest positive, active behavior while the other hints at negative, constraining activities. Distributed networked operations will require a language of rules every bit as rich as this English language usage.

The following sections introduce different types and uses of rules that have emerged during preliminary research into distributed networked systems. This list is not meant to be inclusive. In many ways it corresponds to the simple dictionary list above: a list that should serve as inspiration to the discovery of more types of rules for distributed networked operations.

Rules for Directed Action. Rules can direct agents in a system to take fully prescribed actions. Such rules are intended to achieve explicit control of an agent, which suggests that an agent provided only with authorized actions should not take unauthorized actions. In a complex environment, rules for directed action should only be developed as general rules, since specialized rules by definition require a simple environment. Otherwise, information conditions may arise for which specialized rules result in unintended, perhaps detrimental, directed action. In other words, the precise information conditions cannot be guaranteed from one use of the rules to the next, so only general rules will behave as expected over many different conditions.

Rules for Forbidden Behavior. This type of rule is the obverse of rules for directed action. Rules for forbidden behavior prevent actions from occurring and should also be applied with the same caution as rules for directed action. Otherwise, these rules may actually cause undesirable inaction.

Rules for Conditional Responses. A different kind of rule invokes conditional responses to the environment. These "if-then" type statements have a long pedigree in computer programming. They will be very useful in distributed networked operations since operational rules will be digitally recreated in automated systems and unmanned collectives. As anyone who has written computer code can attest, however, nested conditional statements can be extremely difficult to de-conflict and debug, and they frequently create unexpected pathological behavior. This is because the statements can be invoked in an extraordinarily large number of ways, some of which cannot be identified until the systems interact with the real world. To clarify the difference between this type of rule and the previous two, rules for directed action are rules that mandate "always do this," rules for forbidden actions mandate "never do this" and rules for conditional response mandate "if these conditions are met, do this."

Conflicting Rules. As more rules are added to a rule set, there is an increasing potential that the rules will conflict with each other (particularly if rules are nested or mutually dependent). As the number of rules grows and the number of agents increases, it becomes nearly impossible to create a completely de-conflicted rule set. In complex environments this can only be avoided by keeping rules few and general.

General Rules v. Special Rules. Sometimes, however, it is impossible to keep rules few and general; both general and specialized rules may be required to coexist in the same rule set. In complex environments, general rules are applied at the coarse scale based on coarse scale information and specialized rules are triggered at local scales based on fine scale information. This means that general rules will be invoked more often than specialized rules. Specialized rules will therefore more often conflict with general rules than the other way around. The extent to which specialized rules are allowed to conflict with general rules and the character of allowable conflicts are an indication of the freedom of action enjoyed by agents in a system. This that some rule conflicts can be beneficial to system performance.

Rule Applicability. There are cases in which not all rules apply equally to all elements in a system. This is known as the "Red Lily Effect."[24] This effect occurs because rules are always crafted without enough understanding of the potential future states of a system. Even general rules can eventually become irrelevant to some population of the elements. Although the larger system might still comport to the original intent of the rules, some elements would invoke the rules and some would not. It is more difficult to trace causality and assess the impact of particular variables when rules are irregularly applied.

Rule by Exception. There are cases in which rules should be invoked by exception, meaning that a rule might be required to create very narrowly defined, specialized behaviors that occur in very rare cases. When such rules are invoked, it is usually to ensure a rarely occurring event is noticed and the anomaly that caused the event is addressed. In this case, the rule is usually invoked when all other authoritative, forbidden, or conditional rules are in force and the anomaly arises. This type of rule can be used when a system is performing relatively well, but manual intervention can achieve better performance in certain circumstances.

"Dutch Boy" Rules. There are also cases in which a system operating according to some rule set has very good performance over a large number of different environments yet performs poorly in a small number of environments. In these cases, it may be wiser to create special rules that imperceptibly improve average performance but prevent the

instances of poor performance. These are called "Dutch Boy" rules, after the fictional character who prevented a flood by putting his finger in a hole in a dyke. By making a tiny local correction, the Dutch Boy prevented a huge, global catastrophe.

Over-constraining Rule Sets. It is possible to over-constrain a system with rule sets that are too restrictive. This can occur when only a relatively small subset of all possible configurations of the system satisfies the rule set. To clarify, these are not necessarily conflicting rules, but a set of rules that is internally consistent yet not robust. Over-constraining rule sets result from specialization and inhibit adaptive behavior in a system.

This chapter has discussed adaptive command and control by presenting a general theory of adaptive control in complex environments and by showing how the theory is applied through the creation and use of rule sets. It should be clear to the reader that the development of rule sets for adaptive command and control is far from trivial. Operators and engineers should approach this task with a fair dose of humility and expect substantial trial and error in experimentation and testing. Indeed, this trial and error is exactly the kind of feedback that will allow the defense community to learn and adapt what developing distributed networked forces.

[1] The next two sections originally appeared in Jeffrey R. Cares and CAPT Linda Lewandowski, USN, "Sense and Respond Logistics: The Logic of Demand Networks," undated, unpublished US Government white paper, 2002, for a discussion of this concept.
[2] Jeffrey R. Cares, Raymond J. Christian, and Robert C. Manke, Fundamentals of Distributed, Networked Military Forces and the Engineering of Distributed Systems, NUWC-NPT Technical Report 11,366, 9 May 2002, NUWC Division Newport, 1.
[3] This thesis is explored in Robert Axelrod, *The Complexity of Cooperation: Agent-Based Models of Competition and Collaboration*, (Princeton University Press, Princeton, NJ, 1997).
[4] True noise is the equivalent in military lexicon to classic Clausewitzian "fog and friction." See Barry D. Watts, *Clausewitzian Friction and Future War*, McNair Paper Number 52, October 1996 at http://www.ndu.edu/inss/macnair/mcnair52/m52cont.html, accessed 30 March 2004. Chapter 7, "The Inaccessibility of Critical Information," is particularly pertinent to control in complex systems.
[5] See Seth Lloyd, "Learning How to Control Complex Systems," SFI Bulletin, Spring 1995, at http://www.santafe.edu/sfi/publications/Bulletins/bulletin-spr95/10control.html, accessed 30 March 2004. Haeckel, Ch. 5, directly connects Lloyd's theory of control to sense and respond system.
[6] http://www.math.psu.edu/gunesch/entropy.html, accessed 30 March 2004, is a web portal containing a host of links to scientific web pages discussing the various uses of entropy.
[7] Claude E. Shannon, "A Mathematical Theory of Communication," The Bell System Technical Journal, Vol. 27, pp.379-423, 623-656, July, October 1948, is still the best reference for this concept.
[8] See Chaitin, The *Limits of Mathematics: A Course on Information Theory and the Limits of Formal Reasoning*
[9] Both these cases are mathematically derived in W.H. Zurek, "Algorithmic Information Content, Church-Turing Thesis, Physical Entropy, and Maxwell's Demon," Zurek, ed., *Complexity, Entropy and*

the Physics of Information, (Addison-Wesley Publishing Company, New York, 1991). Figures 4.1 and 4.2 are reproductions from that text.

[10] This occurrence has a mathematical equivalent in Gödel's Incompleteness Theorem. See Chaitin, *The Limits of Mathematics: A Course on Information Theory and the Limits of Formal Reasoning*.

[11] For a formal proof of this statement and a complete treatment of the concepts of randomness, sense-making and Information Theory, see Gregory J. Chaitin, *The Limits of Mathematics: A Course on Information Theory and the Limits of Formal Reasoning*, (Springer-Verlag, New York, 1998), *The Unknowable (Springer Series in Discrete Mathematics and Theoretical Computer Science,)* (Springer-Verlag, New York, 1999) and *Exploring Randomness (Discrete Mathematics and Theoretical Computer Science)*, (Springer-Verlag, New York, 2001). For an application of these principles to distributed systems, see Jon Barwise and Jerry Seligman, *Information Flow: The Logic of Distributed Systems*, (Cambridge University Press, New York, 1997). For a discussion of how these concepts are applied to observation, cognition and decision, see Tor Nørretranders, *The User Illusion*, (Penguin Books, New York, 1998). Further exploration of the information content of physical systems is found in Wojciech Zurek, ed., *Complexity Entropy and the Physics of Information*, (Addison-Wesley Publishing Company, New York, 1991).

[12] Seth Lloyd, "Learning How to Control Complex Systems".

[13] The remaining sections of this chapter originally appeared in Jeffrey R. Cares, "Rule Sets for Sense and Respond Logistics: The Logic of Demand Networks, unpublished US Government White Paper, 30 March 2004.

[14] Juergen Jost, "External and Internal Complexity of Complex Adaptive Systems," Santa Fe Institute Working Paper 2003-12-070, http://www.santafe.edu/sfi/publications/wplist/2003, accessed 30 March 2004.

[15] Juergen Jost, Santa Fe Institute Working Paper 2003-12-070. Similar themes are explored in Feldman and Crutchfield, "Structural Information in Two-Dimensional Patterns: Entropy Convergence and Excess Entropy," Santa Fe Institute Working Paper 02-12-065, http://www.santafe.edu/sfi/publications/wplist/2002, accessed 30 March 2004; Fatihcan Atay and Juergen Jost, "On the Emergence of Complex Systems on the Basis of the Coordination of Complex Behaviors and their Elements," unpublished working paper, dtd. November 5, 2003; and Carlos Gershenson and Francis Heylighen, "How Can We Think the Complex," unpublished, undated.
[16] Jost, "External and Internal Complexity of Complex Adaptive Systems," 3-4.
[17] Jost, "External and Internal Complexity of Complex Adaptive Systems," 5.
[18] Ricard V. sole, et al., "Self-Organized Instability in Complex Ecosystems," Phil. Trans. Royal Soc. Series B, Special Issue: The Biosphere as a Complex Adaptive System, 2002.
[19] The following discussion is adapted from Manoj Gambhir, Stephen Guerin, Daniel Kunkle and Richard Harris, "Measures of Work in Artificial Life," and Stephen Guerin and Daniel Kunkle, "Emergence of Constraint in Self-Organizing Systems," white papers, http://www.refish.com, accessed 30 March 2004.
[20] As an operational example, negative information in subsurface warfare is still information. Lack of information on a target—knowing where the target is not—is a kind of useful information that is profitably applied to military searches.
[21] Cares and Lewandowski, 2002.
[22] Webster's New Universal Unabridged Dictionary, (Dorset and Baber, New York, 1983).
[23] "Ruleset" as a compound word is not in the dictionary.
[24] After Anatole France's farce by that title in which nobles in pre-revolutionary France are bemused by a law that prohibits sleeping under the bridges in Paris.

5

An Information Age Combat Model

This book has addressed environmental complexity, adaptive processes and the role of rule sets in adaptive operations. The reader might get the impression that enough is known about these topics to produce military hardware for distributed networked operations. There is still, however, a gulf between a philosophical understanding of adaptation and the engineering prowess to make purposeful, stable and controllable adaptation a reality in the battlespace. Chapter 2 cited one of the obstacles to bridging this gulf, the lack of an acceptable model of combat in the Information Age. An Information Age Combat Model is the first step toward developing a value proposition against which technical invention, system testing and program acquisition can be gauged.

This chapter will develop an Information Age Combat Model that satisfies the requirements set forth in Chapter 2 for a transformation in combat modeling: it explicitly represents interdependencies, appropriately captures fine-scale tactical arrangements and can reproduce tipping point behaviors. The model also produces a set of approximate thumb rules to guide Information Age concept development, systems engineering, operational experimentation and program analysis. Since an Information Age Combat Model must explicitly address networks, the following three sections will formally introduce the topic of networks and then describe and develop the basic structure of the proposed model.

MATHEMATICAL STRUCTURE OF NETWORKS

The term "network" has become a ubiquitous synonym for any connected system; other synonyms such as "grid," "chain" or "mesh" are likewise creeping into operational language. Very few who use these words exhibit understanding that the terms have very specific definitions in mathematical Network Theory. A "grid," for example, is technically a "lattice of degree four," which means there are exactly four links connected to each node. There are no shortcuts in a grid so information and other commodities take too long to move throughout the network. Grids also have a very rigid structure so they are not adaptive. These properties make the grid a poor candidate structure for distributed networked operations. While the use of the term grid suggests that it is intended as a metaphor rather than a technical requirement, the fact that it is not even a good metaphor seems to be missed altogether.

This is not an arcane point—one should care about the specific mathematical properties of a network for two very practical reasons. The first reason is that different networks have dramatically different properties. Selecting a network type simply because a phrase has popular appeal can result in systems with the wrong properties for the tasks they are required to perform. Second, many of the characteristics which concept developers ascribe to new operational concepts, such as "adaptation," "self-synchronization," "networked effects" or "robustness," have specific mathematical definitions that can be derived from the science of networks. Any model of distributed networked combat that ignores the mathematical properties of networks would therefore inappropriately represent Information Age combat.

Networks are best understood from three main perspectives:

- Network structure—the definition of links, nodes and connection rules

- Network dynamics—the mechanisms by which networked effects are achieved
- Network evolution—the behavior of the network as it adapts in a competitive environment

The Information Age Combat Model will be presented from each of these perspectives.

NETWORK STRUCTURE

The Information Age Combat Model should have the mathematical structure of a network, which at the most basic level is a collection of *nodes* connected by *links*. A taxonomy is required to differentiate between the kinds of links, nodes and connection rules that the model would have. Nodes are defined as elements in a process that are *sensors, deciders, influencers*, or *targets*. The following definitions comprise the basic Information Age Combat Model taxonomy:

- Sensors receive signals about observable phenomena from other nodes and send them to deciders
- Deciders receive information from sensors and make decisions about the present and future arrangement of other nodes
- Influencers receive direction from deciders and interact with other nodes to affect the state of those nodes
- Targets are nodes that have military value but are not sensors, deciders or influencers

These definitions require a few clarifications. First, nodes can have a characteristic called "side" (e.g., friend, foe, neutral). Second, in some taxonomies, targets always belong to an adversary. In the Information Age Combat Model, targets are anything of military value on either side that is not a sensor decider or influencer. Third, sensor logic (determining if a signal is received or not) is not considered a decision-making capability. Sensor logic is therefore contained within sensors. Fourth, all sensor information must pass through a decider; "sensor-to-shooter" is allowed, "sensor-to-bullet" is not. Deciders know the location of their own side's targets, influencers and *disconnected* sensors only if they are detected by own side's sensors.

Nodes are linked to each other by directional connections, that is, links. An example of a link is an observable phenomenon that emanates from a node and is detected by a sensor. In this case, links might be radio frequency (RF) energy, infrared signals, reflected light, communications or acoustic energy. Phenomena detected by sensors are communicated to deciders, the communication constituting another kind of link. Deciders issue orders to influencers, sensors and targets. Influencers interact with other nodes, typically in an effort to destroy or render those nodes useless. These orders and interactions are also links. Links are not necessarily IT connections between nodes, but represent something more functional; most of the links in the Information Age Combat Model represent tactically driven, operational interactions between nodes.

COMBAT NETWORKS

The links and nodes as defined above constitute a *combat network.* Figure 5.1 graphically represents the most basic combat network, while Figure 5.2 represents a two-sided system. Note that the nodes are color-coded to denote sides: black for friendly and light grey for enemy. Note also the use of various line styles to highlight the different kinds of links in the Information Age Combat Model. These styles will be omitted for simplicity in later figures. This simplification is helpful for the reader at the current stage of model maturity, but it masks the fact that the model will not be useful for practical analysis until more sophisticated values are assigned to links. This is not a trivial embellishment and should be the next major advance in the development of the Information Age Combat Model.

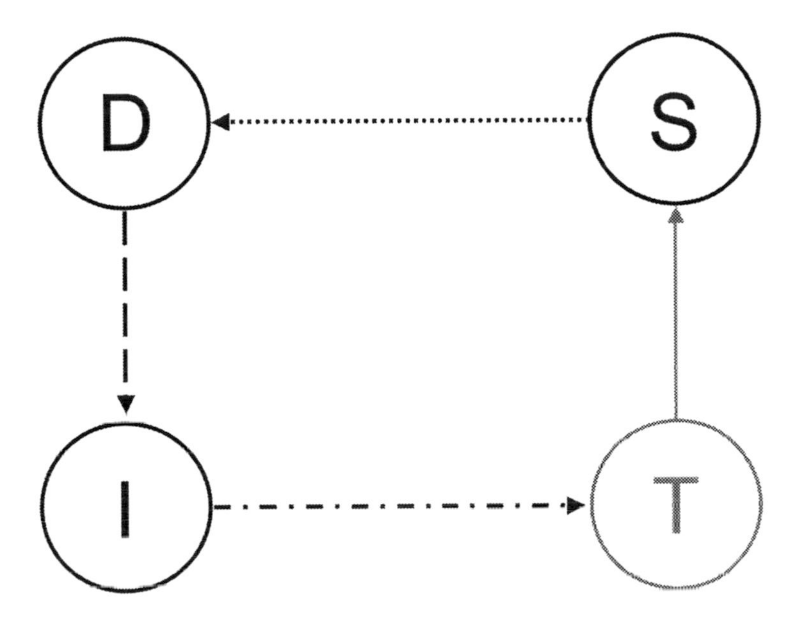

Figure 5.1—Simplest Combat Network

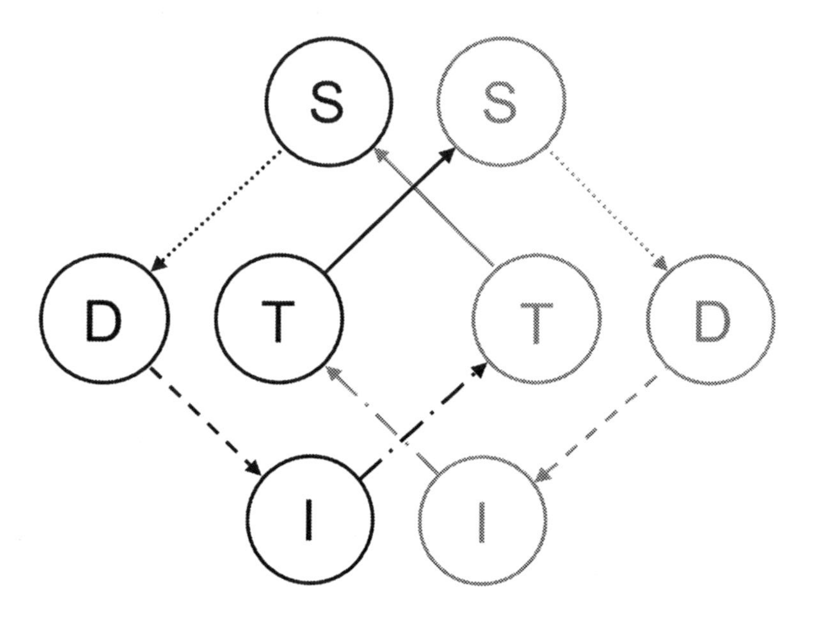

Figure 5.2—Simplest Two-Sided Combat Network

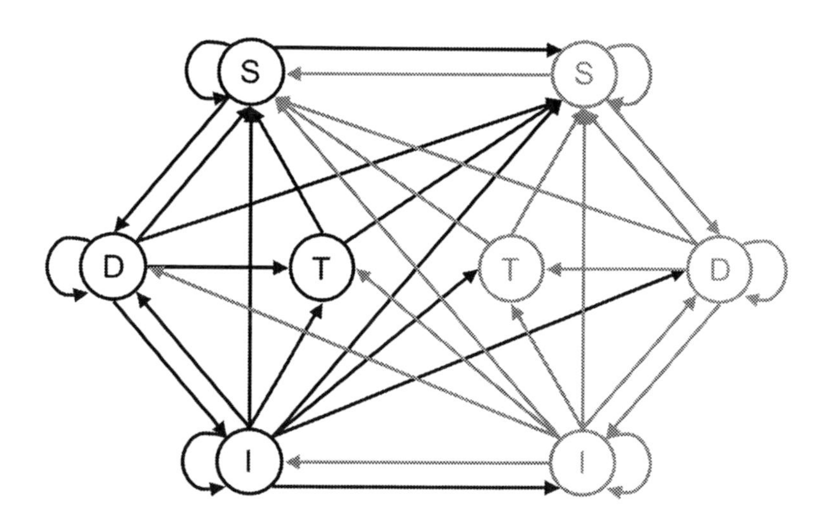

Figure 5.3—Simplest Complete Combat Network

DIMENSIONS AND COMPLEXITY

Figure 5.3 represents the *simplest complete* combat network that can be created from the taxonomy, definitions and assumptions in the previous two sections. This diagram represents all the ways in which sensors, deciders, influencers and targets interact meaningfully with each other. The two dimensional surface of this paper obscures the inherent complexity of this "simple" network: there are at least 36 different dimensions in which this network operates. Dimensionality is easier to calculate from a different type of network representation, an *adjacency matrix.* The adjacency matrix in Figure 5.4 is an equivalent representation of the network in Figure 5.3. A "1" in the matrix indicates that there is a link from node listed at the head of the row to the node listed at the head of the column. A "0" indicates that there is no link between those nodes. Note that the connections are directional from rows to columns. For example, the friendly influencer, I, has a link from a friendly decider, D, and the enemy influencer, I, but not from its own side's sensor, S, or target, T, or from the enemy sensor, decision node or target. Counting up all the matrix entries filled with a "1" provides the dimensionality of the simplest, complete combat network, 36. Recall that this is the simplest complete model; one could include many more targets, sensors, decision nodes, and influencers as well as different kinds of links. There can even be more than one link between two nodes. For example, a link from a friendly influencer to an enemy sensor can represent detection of the influencer. There could also be a link between the two nodes representing an attack on the enemy sensor by the friendly influencer.

This high-dimensional structure is complex in the sense that there are an extraordinary large number of different sub-networks that can be created from the simplest complete combat network. In general, the number of different sub-networks that can be created from a directional N x N matrix is $2^{(N2)}$. This number gets very large even for small values of N. If every node in Figure 5.3 were allowed a direct link to every other node, even the "simplest" complete combat network would have 2^{64} alternative tactical arrangements (about 18.4 billion *billions*). Although there is some relief since the Information Age Combat Model does not permit connection of every link between the eight nodes of the simplest complete combat network, the complexity of combat networks will still be quite high since most concepts for distributed networked operations propose tens, hundreds and even thousands of nodes.

		Friendly				Enemy			
		S	D	I	T	S	D	I	T
Friendly	S	1	1	0	0	1	0	0	0
	D	1	1	1	1	1	0	0	0
	I	1	1	1	1	1	1	1	1
	T	1	0	0	0	1	0	0	0
Enemy	S	1	0	0	0	1	1	0	0
	D	1	0	0	0	1	1	1	1
	I	1	1	1	1	1	1	1	1
	T	1	0	0	0	1	0	0	0

row maps directionally to column = 1, 0 otherwise

Figure 5.4—Adjacency Matrix

Figure 5.5, a plot of $2^{(N2)}$, puts combat network complexity in perspective. Combat networks with more than 17 nodes can contain more sub-networks then there are particles of matter in the known universe (Ω, in Figure 5.5). A plot of N!, a number that also gets quite large quickly,[2] looks almost linear when compared to $2^{(N2)}$. Trying to find the best arrangements of nodes and links in this huge space of possibilities would be extraordinarily exhaustive. This is one of the first conclusions one can draw from the Information Age Combat Model: because of the extreme complexity of networked combat, fine-scale control in distributed networked operations should not be centrally dictated. Decentralized control—local tactical arrangements by local commanders capitalizing on locally perceived advantage—avoids addressing all the complexity at once and provides the best command and control for complex combat networks.

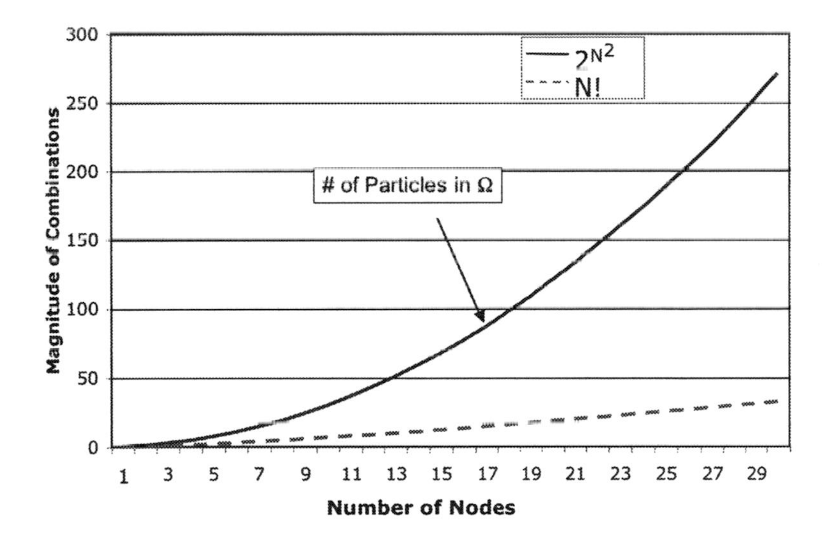

Figure 5.5—Network Dimensionality

COMPARISON WITH EXISTING MODELS

One feature of network representations is that two or more nodes can be *contracted* so that the functions or values of a group of nodes can be represented by a smaller number of nodes. A single node can therefore contain the attributes of, say, a sensor, a decider and a target. Contracting a collection of sensors, deciders, influencers and targets into a set of nodes with one sensor, decider, influencer and target in each node leads to an interesting result that approximates Lanchester equations. This is an additional benefit of using a network mathematics approach: if traditional models like Lanchester Equations can be represented using this framework, then traditional warfare and distributed networked warfare can be compared using the same model. It is impossible to make this comparison with existing models. Figure 5.6 shows what a "Lanchester Network" would look like.

Figure 5.6—Lanchester Network

NETWORK DYNAMICS

Advantageous behaviors occur in local tactical arrangements because of the dynamic interaction between nodes over links. Specific arrangements of links and nodes that create combat value are cycles, sub-networks in which the functions of nodes flow into each other over a path that revisits at least one node once. If there are no cycles in a network, then no useful networked function occurs; advantage in distributed networked operations arises only from these dynamic, often *autocatalytic* (compounding) cycles. Current Network Centric Warfare literature and contemporary combat models do not adequately describe these effects.

Although network theory allows for one-or two-node cycles, such cycles (known as "1-cycles" and "2-cycles") are ineffective in networked combat. A single sensor disconnected from a combat network is a sub-network of the larger set, but has low value until it is connected to the larger combat network. The same is true of a simple target-sensor pairing. 3- and higher-dimension cycles are the source of networked effects in distributed networked operations. There are of four types of these cycles in the Information Age Combat Model.

Types of Cycles

The first type of cycle is a *control cycle* that establishes direct control over a side's assets. Figure 5.7 displays three control cycles. Cycle A is a control cycle where a decider, D, implements direct control over sensor S_1. A second sensor, S_2, observes phenomena from S_1 (perhaps location data about S_1) and reports it to D. D then sends a control signal to adjust the position of S_1, which completes the cycle. Similarly, Cycle B is represents a control cycle in which sensor S receives information from target T and passes that information to decider D, which then sends a control signal to the friendly T. T's movement is detected by S, which completes the cycle. In the third cycle, Cycle C, decider D sends an order to influencer I. Phenomena concerning I's state are detected by S, which communicates back to D. D continues the cycle by sending additional control signals to I.

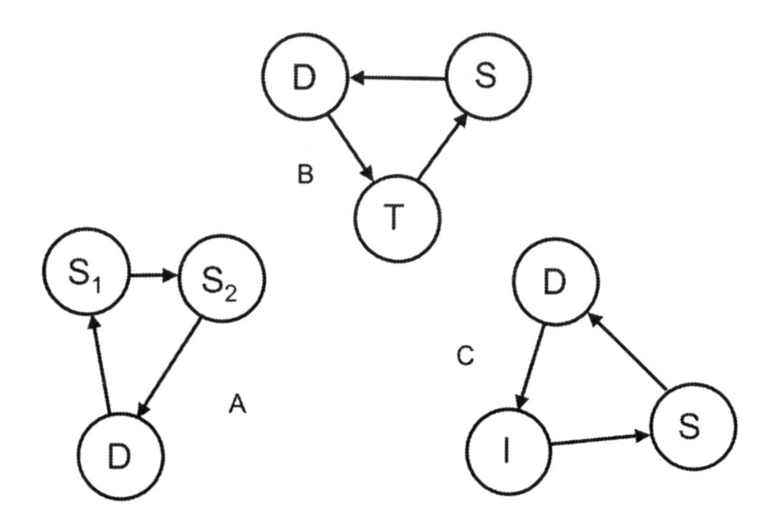

Figure 5.7—Control Cycles

The second type of cycle is a *catalytic control cycle* in which a decider controls assets based on information about friendly assets that are not actively controlled by the decider. Figure 5.8 shows three catalytic control cycles. In catalytic control Cycle A, D controls sensor S_2 while sensor S_1 receives information from both S_2 and target, T, and reports it to D. In this case, D may be seeking, say, to place S_2 farther away from itself than S_1, perhaps to better sense in the vicinity of T. D's decision about S_2 is directly influenced by S_2 but also indirectly influenced, or "catalyzed," by information about T. Similar catalytic control activity can be observed in cycles B and C.

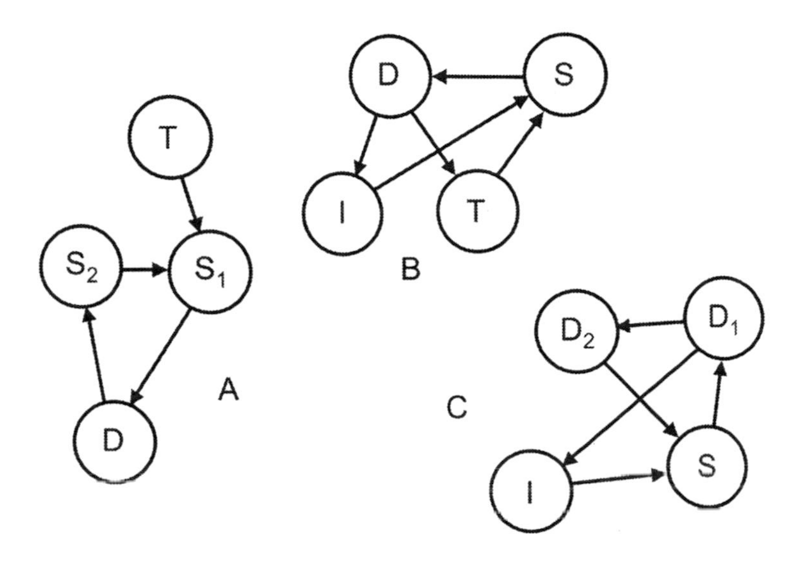

Figure 5.8—Catalytic Control Cycles

The third type of cycle is a *catalytic competitive cycle* that represents control of one side's assets that is catalyzed by information about own side and competing side assets. Figure 5.9 shows two catalytic competitive cycles. In Cycle A, D controls a friendly T and I. A friendly sensor discerns the location of an enemy sensor and a friendly target and influencer, and relays these locations to D. D relocates the friendly T and I based on that information. Movement of the assets is recognized by the friendly S and reported to D, completing the cycle. In example B, similar catalytic control occurs when an enemy D controls an enemy I based on a report by an enemy sensor of the activity of a friendly target, sensor and influencer.

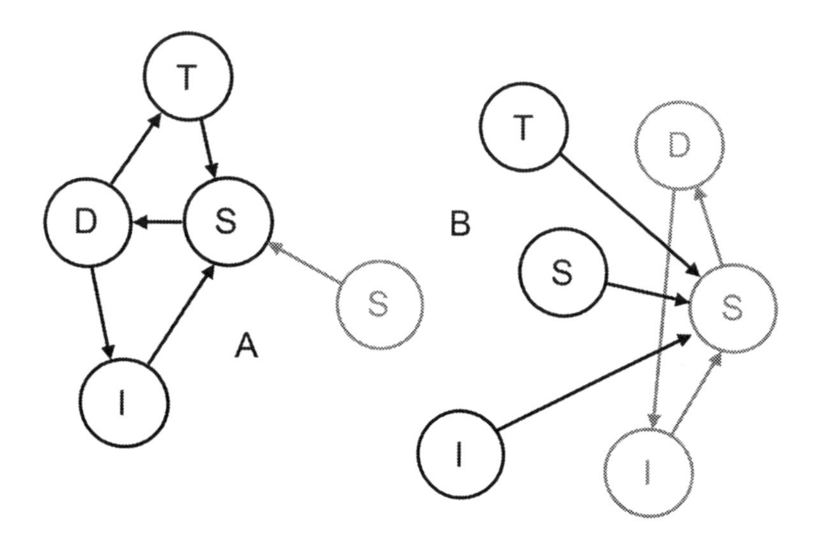

Figure 5.9—Catalytic Competitive Cycles

The fourth type of cycle is a *combat cycle* that represents application of combat power from one side to another (or accidental application from one side to itself). Figure 5.10 portrays two combat cycles. In Cycle A, a friendly sensor detects an enemy sensor. The friendly S relays this information, as well as information on location of friendly T and I to D. D repositions the friendly target and orders the friendly I to attack the enemy S. The attack is observed by the friendly sensor and reported to the friendly D, completing the cycle. In Cycle B, an enemy sensor discerns the locations of friendly target, sensor and influencer nodes and relays this information to the enemy D. The enemy D orders its influencer to attack the friendly influencer. This attack is observed by the enemy sensor and reported to the enemy decider.

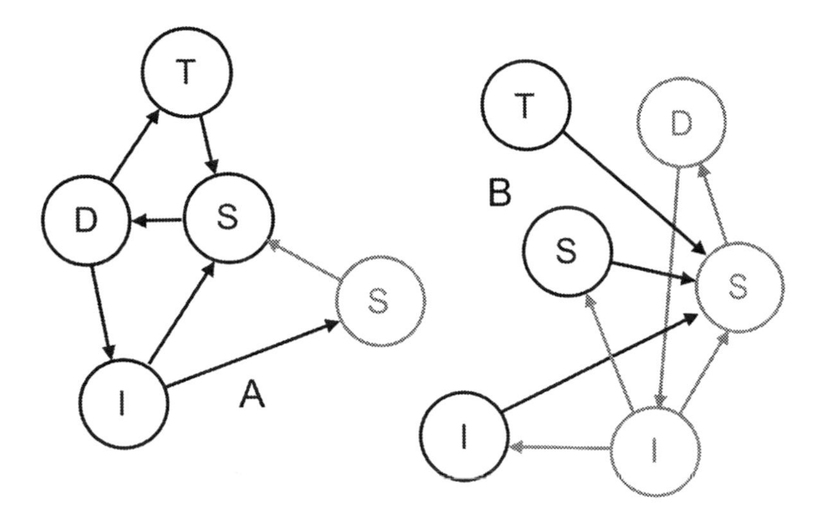

Figure 5.10—Combat Cycles

Measuring Networked Effects

The Information Age Combat Model can be used for more than just diagramming cycles. The model can also use matrix representations to calculate various values and statistics of combat networks. One of these values is the *eigenvalue*, denoted by the Greek symbol λ, a kind of overall matrix value.[3] Because the adjacency matrix representation of the Information Age Combat Model happens to be a "sparse non-negative matrix," the Perron-Frobenius Theorem states that there exists at least one real, non-negative eigenvalue larger than all others. Since adjacency matrix are 1's and 0's, this eigenvalue, the Perron-Frobenius Eigenvalue (PFE), is also guaranteed to have three distinct values which correspond to three measures of networked effects: the absence of a cycle, the presence of a simple cycle and the magnitude of the networked effects.[4]

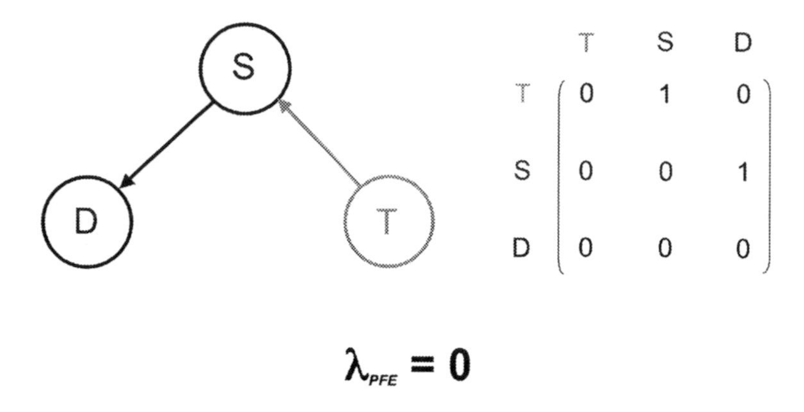

$$\lambda_{PFE} = 0$$

Figure 5.11—Network with No Cycles

The left side of Figure 5.11 shows a network without a cycle; there is no path from a node that returns to that node. The right side of the figure is the adjacency matrix that describes that non-cyclical network. The PFE value for the adjacency matrix is 0. By definition a combat model adjacency matrix with a PFE of 0 represents a network with no cycles (and therefore no networked effects).

Figure 5.12, by contrast, contains a *simple cycle*, a cycle without feedback or feed-forward shortcuts. Without these shortcuts, there can be no compounding, networked effects. The PFE of its adjacency matrix equals exactly 1.0. By definition, an adjacency matrix with a PFE of 1.0 represents a simple network with no networked effects.

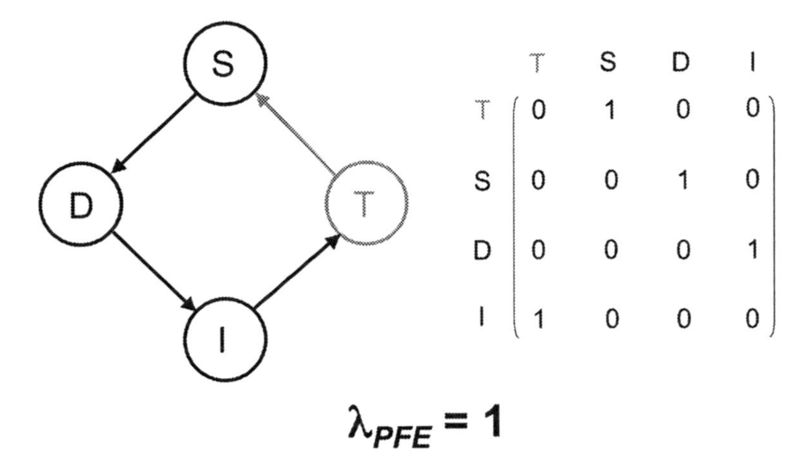

Figure 5.12—Network with a Single, Simple Cycle

Figure 5.13 shows a network with structure over and above that in Figure 5.12. The additional node and links create the feedback and feedforward mechanisms by which networked effects accrue. The PFE of the adjacency matrix representing a structure with such shortcuts is always greater than 1.0 and measures the extent to which these shortcuts compound the simple cycle and create networked effects. These networks are called *autocatalytic sets* (ACSs) because the additional structure compounds effects in a network in the following way: the effects of S_1 and S_2 can both cycle though the network without requiring twice the structure.

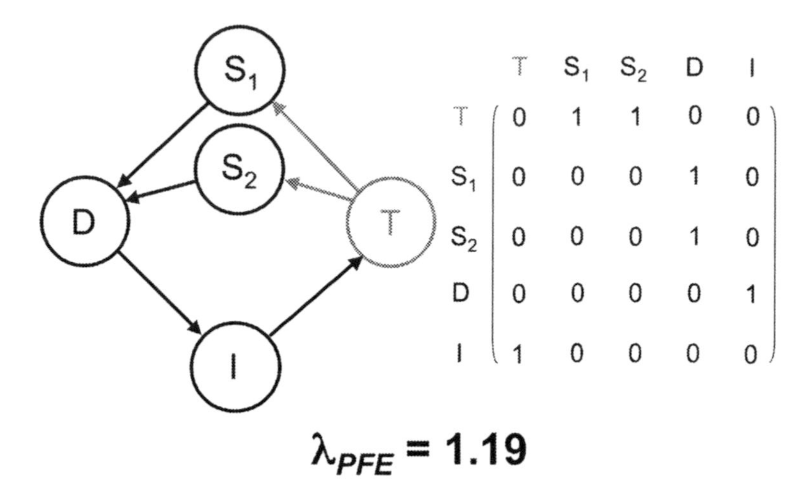

Figure 5.13—Network with an Autocatalytic Set (ACS)

Figure 5.14 shows how the PFE increases with additional linkages, corresponding to an increase in the magnitude of networked effects. However, not all additional links and nodes contribute to networked effects. Figure 5.15 shows how the addition of a link and a node to the basic structure in Figure 5.13 does not change the value of the PFE.

A *core* process is a sub-network that contains all the mechanisms of a structure's networked effects and therefore is responsible for all the value in the PFE. The structure in Figure 5.13 is the core of the network in Figure 5.15. Additional links and nodes that do not add to PFE are called *periphery* links and nodes. It is possible to have more than one core in very large networks.

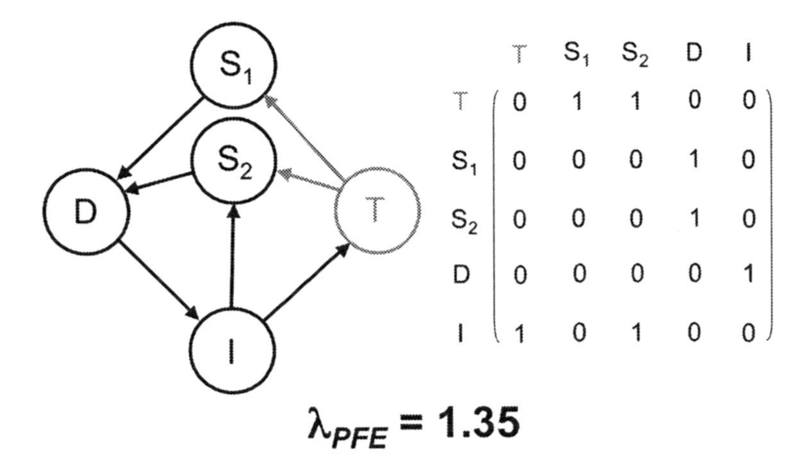

Figure 5.14—ACS with Additional Linkages

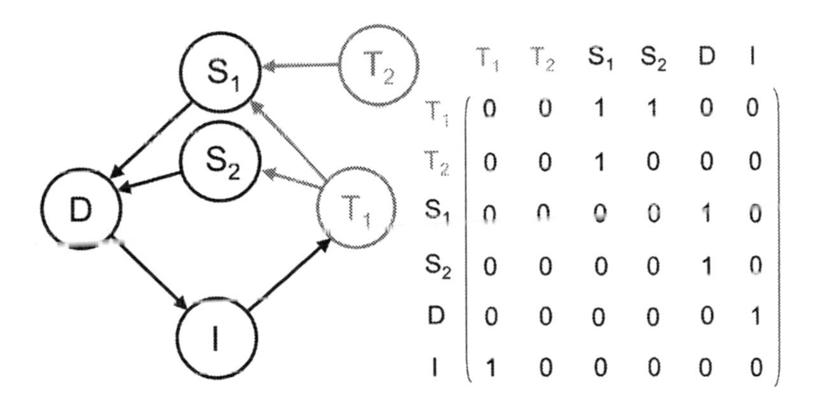

Figure 5.15—ACS with a Peripheral Link and Node

The largest possible λ of an N x N adjacency matrix is N. The networked effects of different size networks can be compared using the ratio PFE/N, which we define as the Coefficient of Networked Effects

(CNE). CNE ranges in value from 1/N to 1.0 (for PFEs greater than 1.0).

Network Evolution

It is evident from these calculations that network performance can increase as networks grow. In fact, networks mature in a way that is quite unlike progressive improvement in most other types of systems. The rest of this chapter discusses the growth dynamics of networks, or *network evolution.* The sections describe how exploitable properties evolve as a complex network grows. A type of rapid connectivity in networked structures will be explored, mechanisms of adaptation and learning will be defined and convergence toward a set of descriptive statistics will be discussed. The potential for using these statistics to quantify combat network performance will also be addressed.

Punctuated Growth in Complex Networks

One of the most important phenomena in network evolution is *punctuated growth*, a pattern of sudden connectivity that occurs as a network grows from collections of small node clusters into a larger, more complex structure. A simple thought experiment demonstrates this rapid growth. Imagine 400 buttons and many pieces of string on a table. Imagine randomly selecting a button and a piece of string, tying them together and returning them to the table. Now imagine repeating this process indefinitely. You will eventually select a button that already has a string tied to it; perhaps you will also select a string that was previously connected to a button. Soon the table will be populated with many small clusters of buttons and strings. At some critical threshold, picking a few additional buttons and strings will connect almost all the small clusters into one large collection of buttons and strings (called the *giant component* of the network of buttons and strings).

Plotting the size of the giant component against the ratio of strings to buttons more exactly describes punctuated growth. The horizontal axis in Figure 5.16 is the ratio of strings and buttons that have been selected at least once. As this ratio approaches 0.5, a giant component of buttons and strings will form and then dramatically increase. The curve flattens after this ratio hits 0.6, however, as each additional string adds only marginally fewer buttons to the network. Obviously, a connected network is not guaranteed by this method (the curve is asymptotic to the maximum number of buttons) but the curve displays a rapid transition from unconnected clusters of nodes to a very large contiguous network.[5] Although the buttons and strings in the thought experiment were randomly connected, complex networks connected by other mechanisms experience the same type of punctuated growth and similar S-shaped curves describe the growth profiles of these networks.

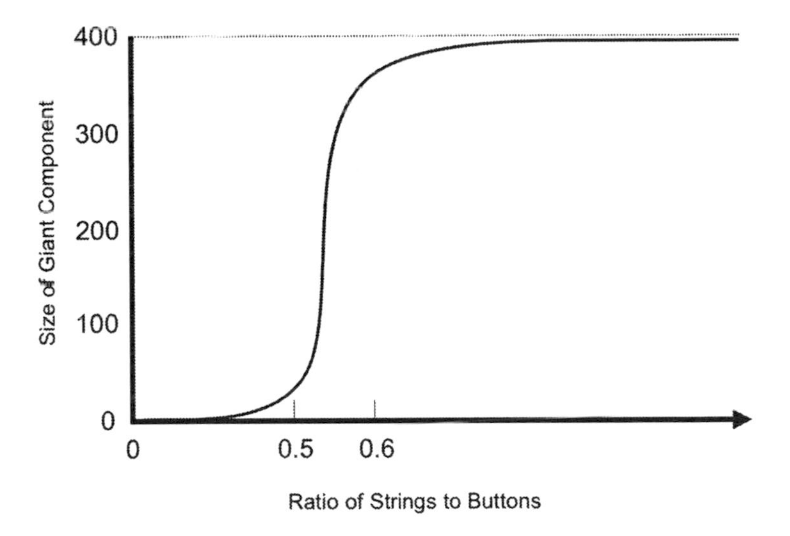

Figure 5.16—Buttons and Strings

There is a very important conclusion one should draw from Figure 5.16. Some have suggested (particularly in reference to distributed networked naval forces) that "numbers count."[6] This is true in Figure 5.16 in the sense that the number of buttons determines the steepness of the curve. The tipping point behavior itself, however, is triggered more by the addition of links than by the addition of nodes. In other words, the greatest marginal increase in the size of the giant component comes from the addition of a few links at the critical threshold. About 90-per cent of the increase in the giant component is the result of connecting about 10-per cent more links. So although numbers do count, links count more.

ADAPTATION IN COMPLEX NETWORKS

Readers familiar with calculus-based engineering might find the curve in Figure 5.16 quite familiar. With this traditional perspective, however, they would mistakenly assume that the important system behaviors are at the "knees" of the curve (where the curve "tips" up and then down) and at the mid-point of the "S." However, there is *latent structure* in the "tail" of the curve to the left of the critical threshold that is far more important than the tipping point itself. This latent structure consists of links that create small clusters of nodes that eventually connect at the tipping point, but the tipping point will not occur unless these latent links are properly configured. One of the mechanisms that create this latent structure is *growth with preferential attachment.*[7] Particularly in a combat network with sensors, the initial links create node clusters that inform the placement and connection of subsequent clusters of links and nodes. As more links and nodes are added, the network evolves from one with no cycles to one with multiple simple cycles, and finally to one with autocatalytic cycles and complex networked effects.[8]

In some networks, latent structure may consist of up to 90-per cent of a network's links and nodes but the giant component remains small. This is an important property that should be exploited in military networks: it is possible for a distributed networked force to be very well connected locally, yet keep its global structure ambiguous to enemy observers. The network can then be rapidly configured for collective action at the time and place of a commander's choosing. The commander can complete the operation and then return to an ambiguous state. Recall the General Theory of Complex Control from Chapter 4 and note that since most of the network will appear pattern-less to an adversary, commanders who operate in this fashion will force an extraordinary sensing and intelligence burden upon their competitor.

The presence of latent structure also helps the network morph smoothly in response to environmental or competitive change. When the environment or competition changes substantially, the arrangement of links and nodes and, therefore, the networked effects, could become irrelevant to the new competition or environment. The network will require reconfiguration to adapt to its new purpose, but because of the presence of latent structure, this adaptive reconfiguration can be achieved with a re-wiring of only 5-to 10-per cent of the links. Herein lies another property that will be very useful for distributed networked operations: commanders can fluidly redirect their main efforts while still obscuring reconfiguration efforts to enemy observers.

Latent structure is sometimes called "neutral" structure because it does not contribute to networked effects until it is incorporated into a re-wiring. A measure of adaptability is the amount of latent structure—the amount of *neutrality*—in a complex network.

CORE SHIFTS IN COMPLEX NETWORKS

Adaptation can change the location of dynamical structures in a combat network. A relocation of cores of networked effects is called a *core shift*. In a core shift the central mechanisms of networked effects move from one subset of links and nodes to another subset of links and nodes. The following example shows a core shift in a combat network that evolves from initial detection of a target to a relocation of influencers in preparation for firing and then to an attack of the target.

Figure 5.17 shows a decision node controlling a group of sensors that detect a target. A box outlines the core responsible for networked effects and shading highlights the corresponding elements in the adjacency matrix. The nodes and links outside the boxes are the two peripheral nodes, I_1 and I_2, and the enemy target node, T.

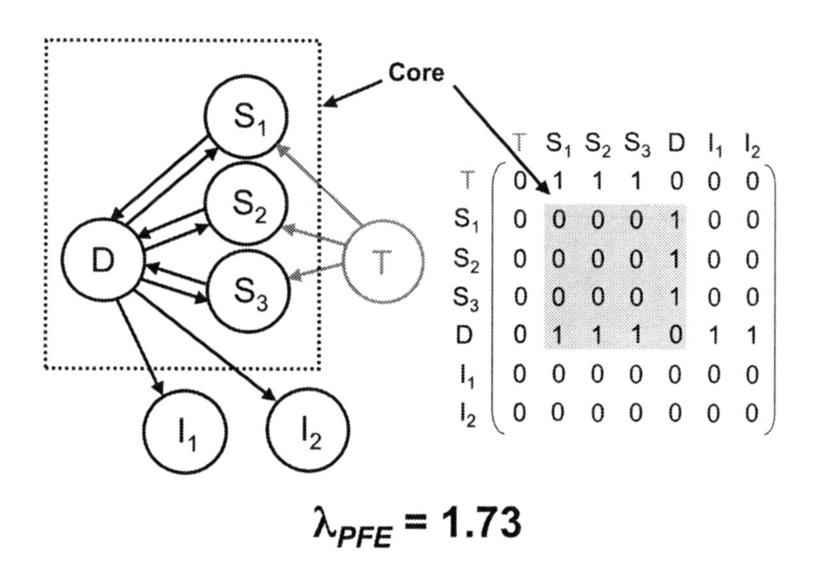

Figure 5.17—A Core of Sensors

Figure 5.18 shows the network's tactical adaptation. One of the sensors tracks the target and the two other sensors monitor relocation of the influencers in preparation for a potential attack. Note that the core has expanded to include the influencers and the PFE has changed as a result. Also note that this was accomplished by removing the links that were used to control two sensors and by the adding new links from the two influencers to two sensors.

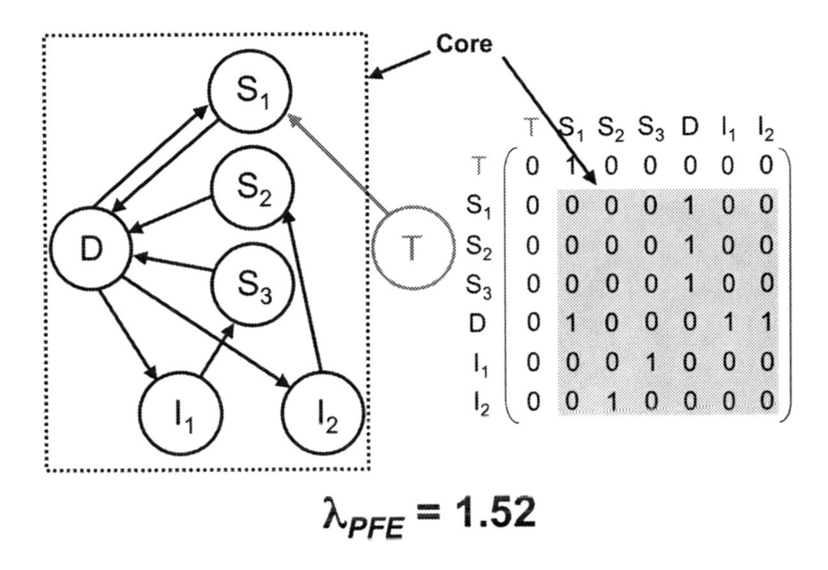

Figure 5.18—A Core of Sensors and Shooters

The network initiates an attack on T in Figure 5.19. The core shifts once again as represented in the lower right corner and the shaded areas of the adjacency matrix. Sensors S_1 and S_2 have been re-allocated to search for additional targets and no longer have a role in the engagement against enemy T. S_1 and S_2 are now peripheral to the network but, most importantly, T is in the core. Again, the PFE has changed with this shift in the core.

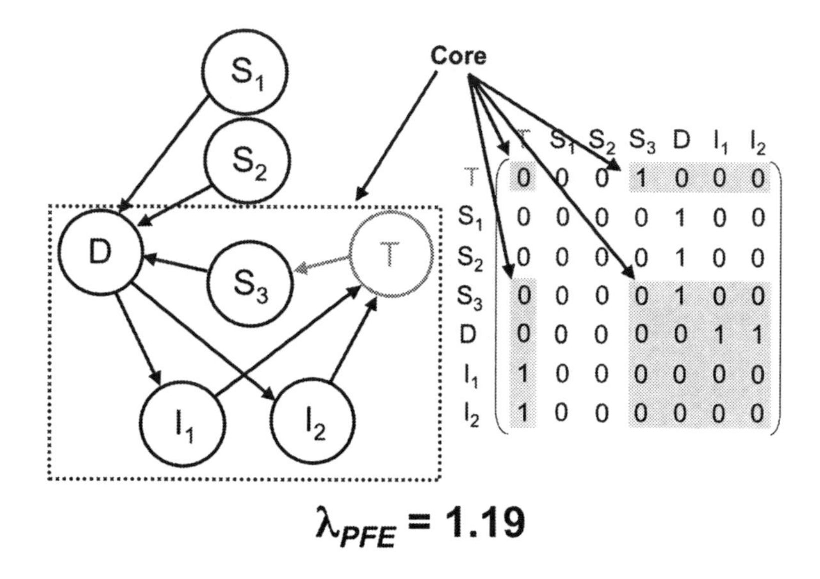

Figure 5.19—Attacking with a Core

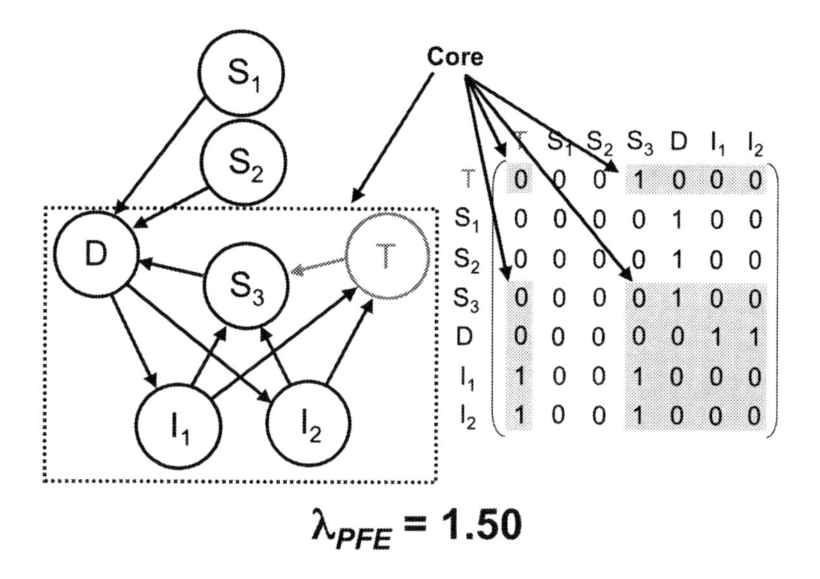

Figure 5.20—Attack by the Core Continues

The attack continues in Figure 5.20. As the influencers engage enemy T, damage to T is monitored by sensor S_3. S_3 communicates this information to the decision node D, which continues to control I_1 and I_2. Note that there is no core shift between the last two figures, just an intensification of networked effects. These additional network interactions are measured by an increase in PFE.

STATISTICAL CONVERGENCE IN MILITARY NETWORKS

A great deal of recent study is providing new insight into the structure, dynamics and evolution of such systems as protein interactions,[9] the internet,[10] the worldwide web,[11] scientific collaborations,[12] ecological food webs,[13] open source software[14] and patterns in motion picture actor employment.[15] The main products of this work are the identification of new classes of network structures and a catalog of statistics describing their characteristics.[16] These statistics measure adaptability, robustness, survivability and many other desirable properties that should be engineered into distributed networked operations. A more complete list of these statistics is contained in the literature, but the most useful measures and their desirable values are listed here. These properties and their suggested values are recommended as thumb rules for Information Age analysis and experimentation. They can each be calculated directly from the Information Age Combat Model.

Number of nodes, N. Although some concepts refer to future single platforms as having network-centric capabilities, networked effects depend on the presence of a large number of nodes. In general, significant networked effects are unlikely to occur in a network of fewer than 50 nodes. The steepness of the connectivity profile in Figure 5.16, for example, is partly a function of the number of buttons: the more buttons, the steeper the curve. As stated above, one of the early, unsubstantiated claims of some network-centric military concepts is that "numbers count." This measurement can evaluate that claim.

Link to node ratio, l/N. NCW literature mandates that all nodes should be directly linked to all other nodes in a maximally connected network.[17] Such networks, however, incur needless overhead. Figure 5.16 shows that very good connectivity can be achieved with orders of magnitude fewer links than the NCW literature suggests. In addition, in networks with link to node ratio lower than the ratio found in maximally connected networks (that is, lower than the ratio of N-1 to 1), the lengths of paths in the network, local cohesion measurements, survivability and adaptability still have desirable values. In fact, all these properties provide advantage when the link to node ratio is approximately 2 to 1. For comparison, the most brittle of structures, chains, have a link to node ratio of approximately 1 to N-1.

Degree distribution. For adaptive network performance, links should not be uniformly distributed throughout a network. A node's *degree* is the number of links connected to it; adaptive, complex networks have a skew degree distribution. A skew distribution means that there are a very small number of highly connected nodes, a moderate number of moderately connected nodes and a very large number of minimally connected nodes. This property is a direct result of cycles in adaptive networks. Skew degree distributions autocatalytically contribute to an ability to reconfigure networked effects with a re-wiring of only about 5- to 10-per cent of the links. Combat networks should have a skew degree distribution.

Size, connectivity of the largest hubs. A skewed degree distribution creates a very small number of very well connected nodes. The largest of these network hubs typically contains fewer than 100 links. Although they are small in number, damage to one can rapidly cascade through a complex network if these hubs are directly connected to each other.

Combat networks should be engineered for so that the largest hubs are not directly connected to each other.

Characteristic Path Length. The characteristic path length is a gross measure of path length in complex networks and is defined as the median (middle ranked value) of the mean of the lengths of all shortest paths in the network. This value grows only by the order of the magnitude of the number of nodes in the network. For example, it takes on average only four links to reach any node from any other node in a network of 10^4 nodes. CPL is a good coarse scale measurement of distance in a network.

Clustering Coefficient. This chapter has presented the mechanism of networked effects in complex networks, autocatalytic feedback. The best type of autocatalytic cycles in a combat network are 3-cycles, because they represent feedback or feed-forward shortcuts in 4-cycle T-S-D-I connections. 3-cycles also contribute to local cohesion, an important tactical principle in military operations. The population distribution of 3-cycles in a complex network is measured by the clustering coefficient, the proportion of a node's direct neighbors that are also direct neighbors of each other. Based on samples from adaptive networks in non-military competitive environments, it is estimated that the average clustering coefficient of a complex network should be between 0.1 and 0.25, meaning that about 10- to 25-per cent of all 3-node collections are arranged in triangular 3-cycles. The distribution of clustering coefficients among all nodes should be skewed, so that a few nodes have a very high clustering coefficient, a moderate number of nodes have a moderate clustering coefficient, and a high number have a low coefficient. This creates the condition that not all nodes in a cluster of mutually supporting nodes interact directly with nodes outside the cluster. A skew distribution of clustering coefficients is found in a complex networks with adaptive hierarchy. Average clustering coefficient is a good coarse scale measure of cohesion and self-synchronization throughout a network. The distribution of clustering coefficients is a good measurement of hierarchy in a networked system.

Betweenness. Betweenness is a measure of a node's importance to network structure. Betweenness measures the proportion of shortest paths that pass through a node (although a node need not be a hub to have high betweenness). Betweenness identifies the bottlenecks in network flow, the highest value nodes in a network and potential pathways for cascading pathological effects or damage. Although nodes with high betweenness can negatively impact network survivability and performance, it is difficult to limit high betweenness yet retain other beneficial characteristics like clustering and skewed degree distribution. Low, uniform betweenness in a grid, for example, is the result of properties that make the grid a poor candidate structure for distributed networked operations. A combat network should limit the number of nodes with high betweenness yet retain good adaptive properties. Combat networks should therefore have a skew distribution of betweenness.[18]

Path Horizon. Path horizon measures the number of nodes on average that a node must interact with for constructive self-synchronization to occur. A path horizon of 1 means that a node must coordinate with all its nearest neighbors for self-synchronization to occur. A path horizon of 2 means that coordination should extend to all the nearest neighbors of a node's nearest neighbors, etc. Research shows that self-synchronization occurs when the path horizon is the logarithm of the number of nodes.[19]

Neutrality Rating. Neutral structure is the additional structure in a complex network over and above the minimum connectivity requirements. A simple chain of links and nodes cannot be a complex network; a complex network, however, can invoke simple chains within it. Complex networks adapt by re-wiring simple chains with links selected out of latent structure. Therefore, subtracting the number of links in a network of size N, N-1, from the number of links, *l*, in a given network of size N, and then normalizing to network size produces the neutrality rating, (*l*—N + 1)/N. The neutrality rating measures adaptability; combat networks should have a neutrality rating of between 0.8 and 1.2.

Coefficient of networked effects (CNE). The coefficient of networked effects measures the amount of cyclic behavior per node and compares the potential for networked effects in networks of different sizes. CNE is the PFE normalized to network size, PFE/*N*. Based on preliminary results from using the Information Age Combat Model for analysis,[20] complex networks should have a CNE between .1 and .25.

Susceptibility. Susceptibility is a measure of the number of links or nodes that can be removed before dynamic structure begins to break down. For example, the curve in Figure 5.16 works both ways: most of a network's connectivity can be lost with the removal of only 5- to 10-per cent of the network's most well connected nodes. This breakdown can be measured in a degradation of the previously listed properties.

Table 5.1 summarizes these thumb rules for analysis and experimentation. Note that since these are approximations inferred from the study of adaptive networks in other domains, it is entirely possible that different values might be observed in military networks. Note also that these are measurements of model topology only. Nodes and links in a real network are not binary connections, but have more sophisticated valuation. Another area for immediate study should be the refinement of combat network thumb rules. Until such time, the existing Information Age Combat Model thumb rules should be considered as a first approximation to a value proposition useful for developing distributed networked systems.

Property	Range	Effect
Number of Nodes, *n*	$n > \sim100$	Networked effects unlikely to occur with $n < 50$
Number of links, *l*	$l < \sim2n$	$l << 2n$, too brittle $l >> 2n$, too much overhead
Degree Distribution	Skewed	Adaptivity, Modularity
Largest Hub	< 100 links	Hub appears, recedes by reconnection 5% of links
Characteristic Path Length	log(*n*)	Short distances even for large networks (e.g., 10^4 nodes → Average Path Length = ~4)
Clustering	Overall: 0.1—0.25 Distribution: Skewed	Hierarchy, Organization
Betweenness	Distribution: Skewed	Highest: Most important nodes, bottlenecks. Cascade Control
Path Horizon	log(*n*)	Self-Synchronization
Coefficient of Networked Effects (CNE)	0.1—0.25	Networked effects per node
Neutrality Rating	0.8—1.2	Increased adaptation; decreased susceptibility
Susceptibility	Low (random removal) High (focused removal)	Hubs should be kept obscure until needed, damage abatement/repair schemes

Table 5.1—Thumb Rules for Analysis and Experimentation

[1] Substantial portions of this chapter (the next eleven sections) are excerpted from Jeffrey R. Cares, "An Information Age Combat Model," Unpublished US DoD White Paper, 30 September 2004.
[2] N! (spoken, N-factorial), is defined as N x (N-1) x (N-2) x…x 1.
[3] See http://mathworld.wolfram.com/Eigenvalue.html, accessed 30 Sep 2004.
[4] Jain, Sanjay and K. Sandeep, "Graph Theory and the Evolution of Autocatalytic Networks," http://arXiv.org/abs/nlin.AO/0210070, accessed 30 Sep 04. As with any multi-variant mathematical problem, there can be more than one eigenvalue that represents the value of a matrix.
[5] Kauffman, *At Home in the Universe*, p. 54-7.
[6] See http://www.defenselink.mil/transformation/cebrowski_ paper _20041216.html, accessed 01 Aug 2005.
[7] See M. E. J. Newman, Clustering and Preferential Attachment in Growing Networks, Santa Fe Institute Working Paper 01-03-021, Santa Fe: Santa Fe Institute, 2001, for a discussion of clustering and preferential attachment in evolving networks.
[8] Jain and Sandeep, 19-22.
[9] Ito, T., Chiba, T., Ozawa, R., Yoshida M., Hattori M., and Sakaki, Y., "A Comprehensive Two-hybrid Analysis to Explore the Yeast Protein Interactions," Proc. Natl. Acad. Sci. USA, 98, 4569-4574 (2001); H. Jeong, Mason, S., Barabasi A.-L. and Oltvai, Z. N., "Lethality and Centrality in Protein Networks," Nature, 407, 41-42 ((2001); Maslov, S. and Sneppen, K., "Specificity and Stability in Topology of Protein Networks, Science, 296, 910-913 (2002); Sole, R. V., and Pastor-Satorras, R., complex Networks in Genomics and Proteomics," in S. Bornholdt and H. G. Shuster (eds.), Handbook of Graphs and Networks, 145-146, Wiley-VCH, Berlin (2003); and Uetz, et. al., "A Comprehensive Analysis of Protein-Protein Interactions in Saccaromyces Cerevisiae," Nature, 403, 623-627 (2000).
[10] Broida, A., and Claffy, K. C., "Internet Topology: Connectivity of IP Graphs," in S. Fahmy and K. Park (eds.) Scalability and Traffic

Control in IP Networks, No. 4526 in Proc. SPIE, 172-187, ISOE, Bellingham, WA (2001); Chen, Q., et. al., "The Origins of Power Laws in Internet Topologies Revisited," in Proceedings of the 21st Annual Joint Conference of the IEEE Computer and Communications Societies, IEEE Computer Society (2002); and Faloutsos, M., Faloutsos, P., and Faloutsos, C., "On Power Law Relationships of the Internet Topology," Computer Communications Review, 29, 251-262 (1999).

[11] Adamic, L. A., "The Small World Web," in Lecture Notes in Computer Science, Vol. 1696, 443-454, Springer, New York (1999); Albert, R., Jeong, H., and Barabas, A.-L., "Diameter of the World Wide Web," Nature, 401, 130-131 (1999); Broder, A., et. al., "Graph Structure in the Web," Computer Networks, 33, 309-320 (2000); Flake, G. W., Lawrence, S. R., Giles, C. L., and Coetzee, F. M., "Self-Organization and Identification of Web Communities," IEEE Computer, 35, 66-71 (2002); Kleinburg, et. al., "The Web as a Graph: Measurements, Models and Methods," in Proceedings of the International Conference on Combinatorics and Computing, No. 1627 in Lecture Notes in Computer Science, 1-18, Springer, Berlin (1999); and Kumar, et. al., "Stochastic Models for the Web Graph," in Proceedings of the 42nd Annual IEEE Symposium on the Foundations of Computer Science, 57-65, IEEE, New York (2000).

[12] Price, D. J. de S., "Networks of Scientific Papers," Science, 149, 510-515 (1965); Redner, S., "How Popular is Your Paper? An Empirical Study of the Citation Distribution," Eur. Phys. J. B, 4, 131-134 (1998); and Selgen, P. O., "The Skewness of Science," J. Amer. Soc. Inform. Sci., 43, 628-638 (1992).

[13] Dunne, J. A., Williams, R. J., and Martinez, N. D., "Network Topology and Species Loss in Food Webs: Robustness Increases with Connectance," Santa Fe Institute Working Paper 02-03-013 (2002) and Martinez, N.D., "Artifacts or Attributes? Effects of Resolution on the Little Rock Lake Food Web", Ecological Monographs, 61, 367-392 (1991).

[14] Solé, R. V. and Valverde, S., "Hierarchical Small Worlds in Software Architecture," Santa Fe Institute Working Paper 03-07-044 (2003).
[15] Adamic, L. A. and Huberman, B. A., "Power Law Distribution of the World Wide Web," Science, 287, 2115 (2000); Amaral, L. A. N., Scala, A., Barthemely, M., and Stanley, H. E., "Classes of Small World Networks," Proc. Natl. Acad. Sci. USA, 97, 11149-11152 (2000); Newman, M. E. J., Strogatz, S. H., and Watts, D. J., "Random Graphs with Arbitrary Degree Distributions and their Applications," Phys. Rev. E, 64, 026118 (2001) and Watts, D. J., and Strogatz, S. H., "Collective Dynamics of 'Small World' Networks," Nature, 393, 440-442 (1998).
[16] Newman, M. E. J., "The Structure and Function of Complex Networks," SIAM Review 45, 167-256 (2003).
[17] Alberts, et al, p. 256.
[18] Stefan Wuchty and Peter F. Stadler, "Centers of Complex Networks," Santa Fe Institute Working Paper, 2002-09-052, September 2002.
[19] Sergi Valverde and Ricard V. Solé, "Internet's Critical Path Horizon," Santa Fe Institute Working Paper, 2004-06-010, June 2004.
[20] Cares, Jeffrey R. and David R. Garvey, "Topological Analysis of LCS Platform and Associated Off-Board Systems Structure and Composition," Unpublished Department of the Navy White Paper, 2004, http://www.dnosim.com/papers.

6

Wolves on the Hunt

This book has defined the complex environments that require distributed networked forces, discussed how adaptation can be used as a means of command and control and presented a fundamental model of networked collectives of distributed combat power. This chapter brings all those elements together in an operational vignette that portrays how a distributed networked force might be employed.[1] Although the following scenario is primarily a naval operation, a similar story can be told of how distributed networked operations might be employed in a ground, air or even a cyberspace campaign.

The scenario begins off the coast of Red, a country thousands of miles from the nearest Blue operating base. This country has long had an adversarial relationship with Blue and its allies, and Blue regularly deploys naval forces to the international waters off Red's shores. Blue currently has a small combatant and a submarine patrolling the Red littoral in a long term effort to gain background intelligence on local conditions such as shipping and commercial air transportation patterns, seasonal variations on atmospheric and bathymetric measurements, bottom-mapping for potential mine warfare operations and Red orders of battle. The Blue vessels conduct these intelligence operations by routinely deploying manned and unmanned air, surface and subsurface vehicles, launching and recovering underwater arrays and comparing locally acquired information with national intelligence products such as satellite imagery. During the many months that these forces are deployed, Blue operators have been deriving and cataloging the best rule sets for employing their distributed networked forces in this environment.

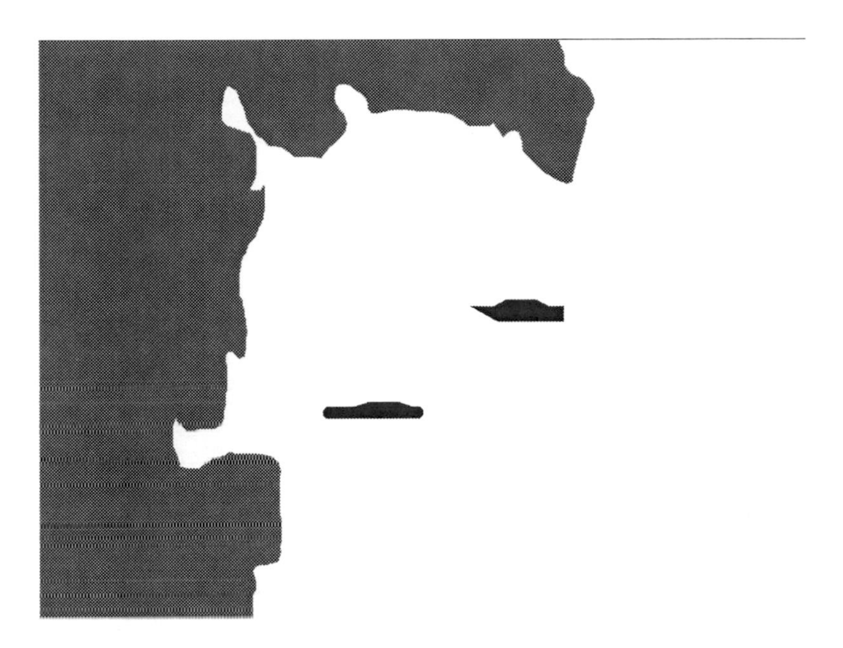

Figure 6.1—Peacetime Operations

Figure 6.1 shows these two vehicles on patrol (the submarine is left of center and the surface combatant is to the right). This figure does not portray the substantial fleet of small fishing boats, coastal merchants and other private craft that would be found off any populated coastline. The commercial airspace and electronic spectrum would be just as cluttered, making Red's littoral a complex environment as defined in Chapter 3.

Over the course of time, Red has become increasingly belligerent toward its neighbor, Purple, a country that Blue and its allies have agreed to defend. Blue's deployed assets have been ordered to increase their intelligence efforts to prepare for a potential deployment of additional Blue forces. The Blue vessels increase the scope of their air operations, establishing a defensive air patrol station (symbolized by the arrow-shaped aircraft icon between two vertical lines in Figure 6.2) and an airborne early warning station (the same symbol underscored by a horizontal figure-eight). After Blue receives intelligence on troubling Red ground force deployments, the Blue force is ordered to send unmanned vehicles into Red air and water space. The smaller submarine, ship and air icons in Figure 6.2 show the deployment of unmanned air, surface and subsurface surveillance assets. Blue operators now invoke rule sets that do not require assets to be refueled or controlled by their original hosts. This increases the on-station time of the off-board assets, decreases the susceptibility of the force to a successful attack on one of the vessels (since sinking one vessel does not destroy the unmanned craft deployed by the stricken vessel) and sufficiently changes Blue operational patterns to confuse Red's intelligence. Blue forces now constitute a core of deployed, networked combat power to which subsequent Blue forces can be connected. Blue forces have shifted their focus from a purely long-term intelligence mission to more near-term efforts to prepare the battlespace for follow-on forces.

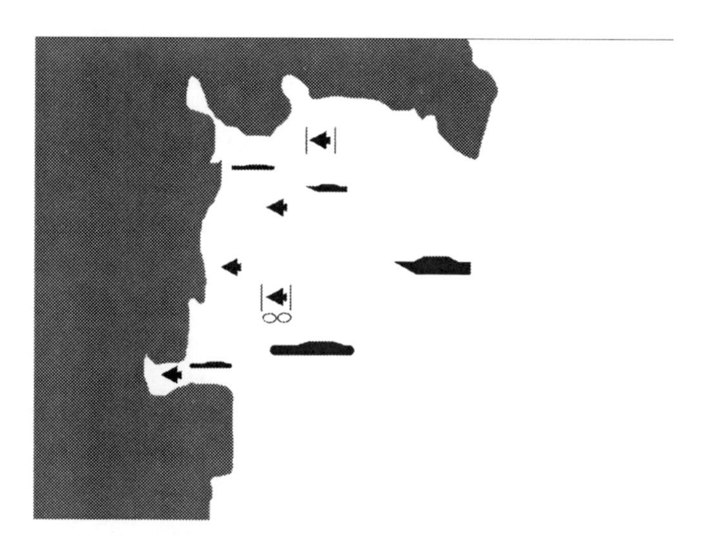

Figure 6.2—Increased Operational Tempo

As the crisis escalates, Blue decides to deploy more assets to the littoral. Blue immediately bolsters its combat power off the coast of Red by sending in more manned and unmanned offboard vehicles. Many of these vehicles are small and lightweight, so they can be deployed at aircraft speeds rather than ship speeds, either as cargo in large air transports (for unmanned air, surface or subsurface assets) or under their own locomotion (for unmanned aircraft). In addition, since the *in situ* Blue vessels are built with a capacity to launch, recover, sustain and control many more assets than they carry in their own hulls, they can immediately assume control of the additional assets as they arrive. Unmanned aircraft can even be launched and deployed well ahead of manned surface combatants that are currently en route to the crisis. Figure 6.3 shows the arrival of additional assets.

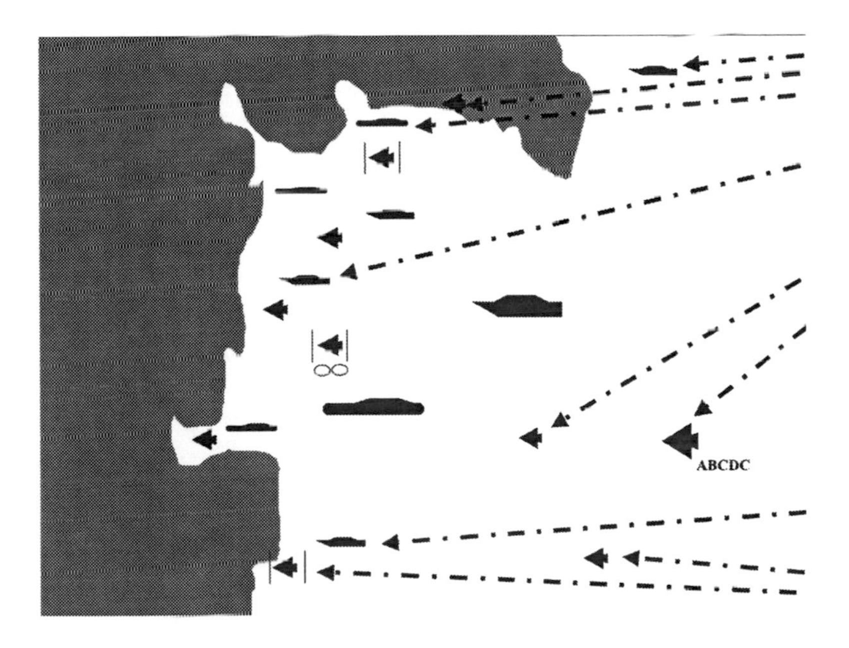

Figure 6.3—Additional Assets Arrive by Air

The Blue units now have an increasingly complex problem to manage. The scope of surveillance and intelligence tasks continue to grow at the same time that local commanders must manage a growing collection of assets in an increasingly challenging competitive space. Blue has decided, therefore, to also deploy an airborne combat direction center (ABCDC, symbolized by the aircraft and acronym in Figure 6.3), manned by a senior officer from a Joint Task Force Commander's staff and a crew of watchstanders. The senior officer in the ABCDC takes overall control of the Blue assets on scene and thereby frees the vessels' operators to focus on tactical, fine scale tasks.

Meanwhile, a main force of Blue vessels, including traditional striking and expeditionary forces, is ordered to proceed at top speed from their positions distributed throughout the theater to the vicinity of Red's littoral. The unmanned aircraft assigned to the striking force are built with interchangeable mission modules that allow them to be configured for battlespace preparation missions (such as undersea and mine warfare), launched well in advance of the main force's arrival and then reconfigured for strike after the main body arrives.

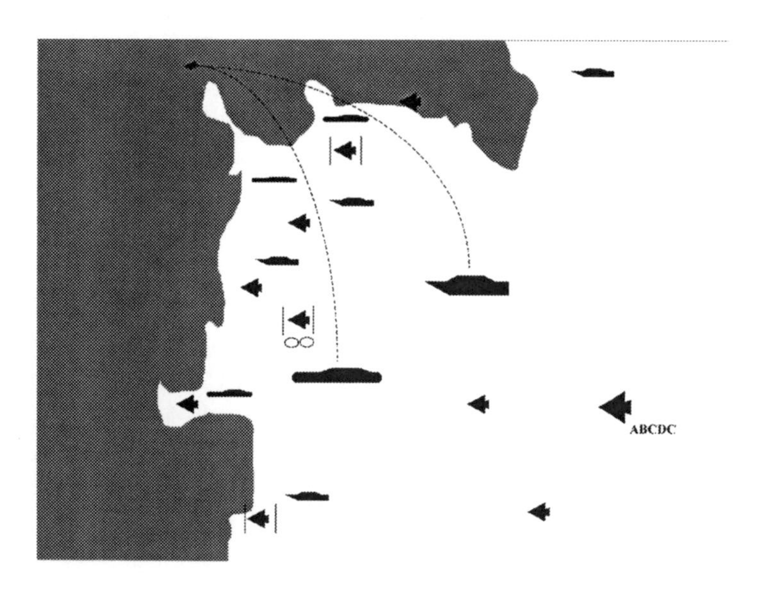

Figure 6.4—Combat Operations Commence

The situation has deteriorated further as Blue receives intelligence that Red is poised to invade Purple and will cross their border at any moment. The Blue vessels are ordered to launch missile strikes on Red mobile command and control nodes as soon as unmanned surveillance assets detect these nodes. The strikes are depicted by the curved lines emanating from the platforms in Figure 6.4.

The large Blue force continues to close on the Red coastline and the local assets continue their intelligence, surveillance and combat operations. Unmanned assets continue to arrive by air transport or under their own power. When the main body striking platforms are close enough, Unmanned Combat Air Vehicles (UCAVs), specifically designed to deploy from the larger platforms and refuel and rearm from the local vessels, are launched to conduct air strikes against Red ground forces. Figure 6.5 shows the arrival of additional assets and a package of UCAVs and manned aircraft striking ground targets in Red.

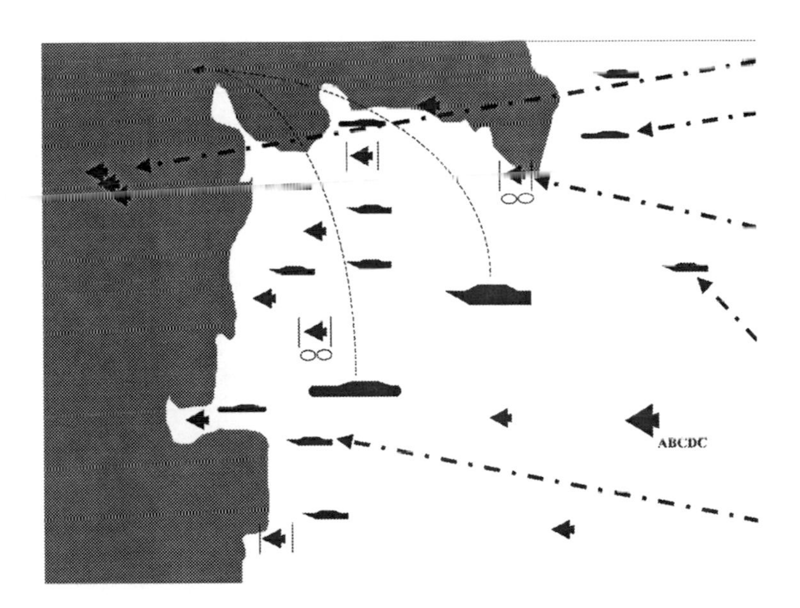

Figure 6.5—Additional Assets Arrive

Chapter 5 described a distributed networked force creating networked effects from feedback cycles in combat networks. Figure 6.6 shows UCAVs cycling between Red targets and the Blue vessels, using the existing sensor and decision-making infrastructure to create cycles of combat power. As the larger forces grow closer, more Blue assets can be connected in a combat network, so more targets can be acquired by more surveillance craft and more air sorties can strike with decreasingly shorter cycles. Figure 6.7 shows how distributed networked operations bring additional assets (including one more manned surface vessel and another airborne combat direction center) to the existing core of combat capabilities to create networked effects.

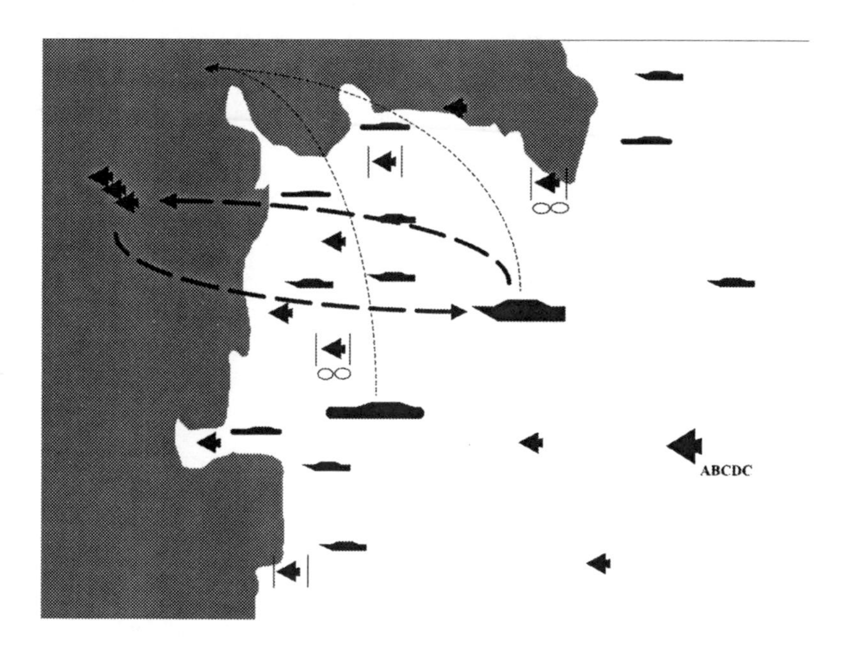

Figure 6.6—UCAV Strike Cycles

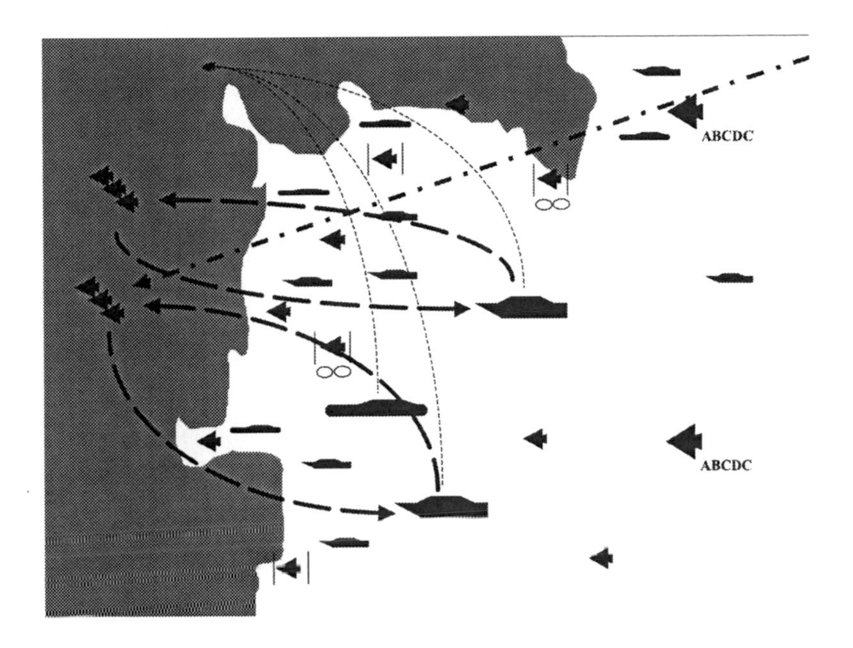

Figure 6.7—Creating Networked Effects

Chapter 5 also described the phenomenon of punctuated growth by which small clusters of links and nodes are rapidly transformed into a large contiguous network. Figure 6.8 shows this occurring in the scenario. As the main body of larger platforms sends more of its assets forward while the main force itself gets closer to Red, a very large component of combat power can suddenly be constituted and applied to Red targets. This also shows how distributed networked operations can transform tactics and Operational Art.

Histories of naval operations describe the changing relationship between major vessels in a fleet and the smaller vessels with which they operate. In the age of sail, the smallest vessels were "packet ships," carrying correspondence between the larger groups of ships or between fleet units and shore bases while acting as scouts en route. Wireless communications arrived not long after steam engines, and navies relieved small ships of their packet duties and transformed them into *screening* units. The screening ships would steam out ahead of the main fleet units to scout for an enemy's main fleet. Once screening ships located an enemy main body, they would harass the enemy units until the friendly main body could either engage or avoid the enemy main body. With the advent of strike and missile combat, the small ships assumed a more defensive role, hugging close to the most important units in the main body to provide anti-aircraft and anti-missile shields.

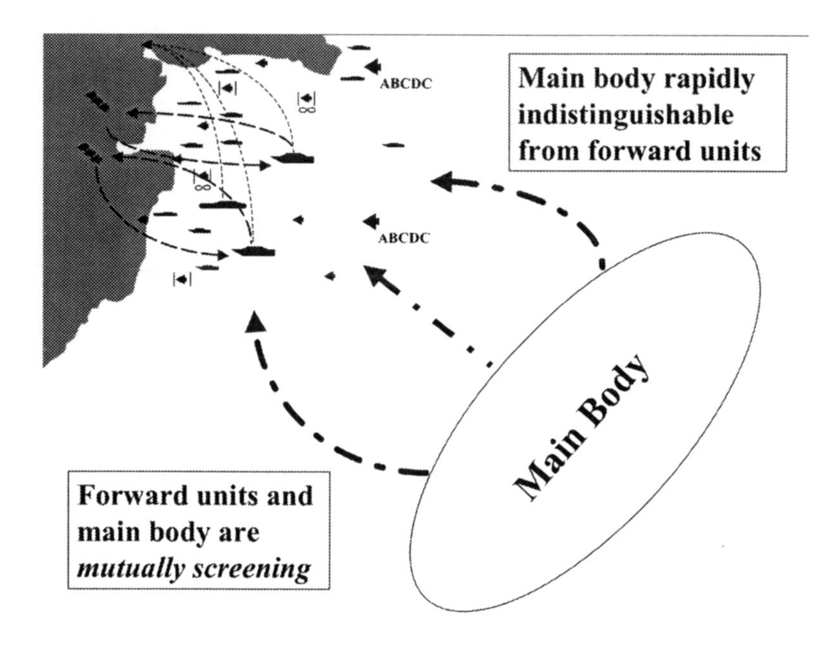

Figure 6.8—Punctuated Growth

Distributed networked operations will allow a fleet's small forward-deployed ships to use the combat power of the main body to screen for the main body itself. This increases the capability and survivability of the forward units, increases options for local commanders and—using the combat power of the main body itself—prepares the littoral for the main force of striking or expeditionary power without putting the main body at risk. The force is therefore *mutually screening*. The distinction between screening units and screened units diminishes as more assets are sent forward and the main body gets closer. The mutually screening force is an additional example of autocatalytic networked effects.

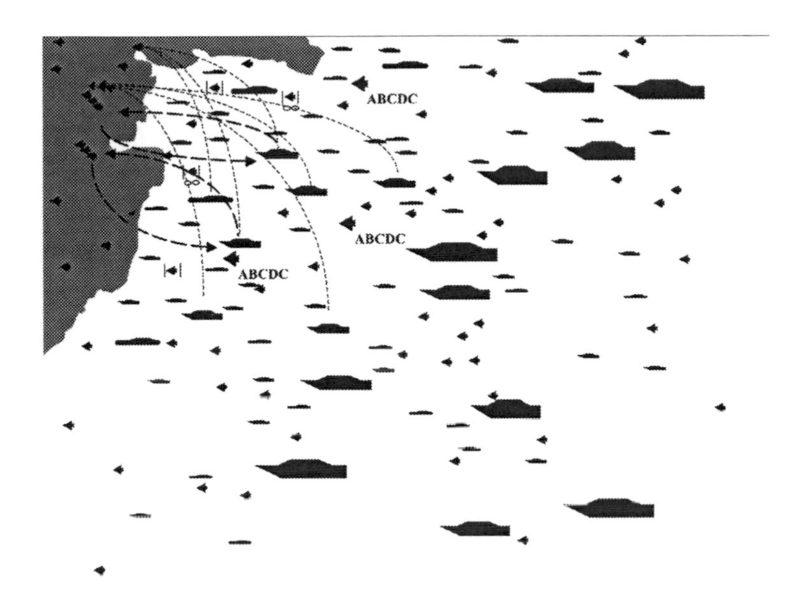

Figure 6.9—A Fully Constituted Distributed Networked Force

Figure 6.9 shows the entire distributed networked force in position off the coast of Red. If this force is properly constituted for adaptive command and control, a mapping of the interactions between and among Blue and Red assets should exhibit the properties found in Table 5.1. This force would have increased survivability, because its assets are dispersed and operations are not crucially dependent upon a small number of high value platforms; increased adaptability, since combat power can be readily exchanged between air, surface or subsurface controlling units; improved balance, since distributed combat power can be massed for both offense and for defense; and improved sustainment in crisis, since combat power is can be sent forward to where sustainment is already deployed (in the hulls of the forward units).

There is a recurring debate in naval force structure analysis about whether it is better to buy a large number of small single purpose ships or invest in a much smaller number of very expensive multi-purpose platforms. When combat power is interchangeable among the units of a distributed networked force, every platform in the force has the potential to be a capital, multi-mission ship. Having multi-mission potential fleet-wide means that a force can be highly reconfigurable and thereby achieve greater economy of force. Distributed networked operations change the "unit of issue" of combat power from the hull to the individual sub-system. This also means that large numbers of ships with only a minimum of organic combat power can be efficiently deployed in peacetime roles yet be rapidly augmented with additional combat power for crisis or war. Rapid augmentation redefines operational speed. The old adage that a force is only as fast as the slowest unit in formation can be revised to reflect the fact that a distributed networked force can now be as fast as the nimblest subsystem. Figure 6.9 compares the transit time for hulls with the transit time for air-delivered platforms in cases typical of naval deployment patterns. Distributed networked operations can couple the speed advantage of air-delivered combat power with the on-station advantages that naval forces provide to redefine operational speed in air-delivered terms. Introduction of high-speed small combatants would compound these benefits even further.

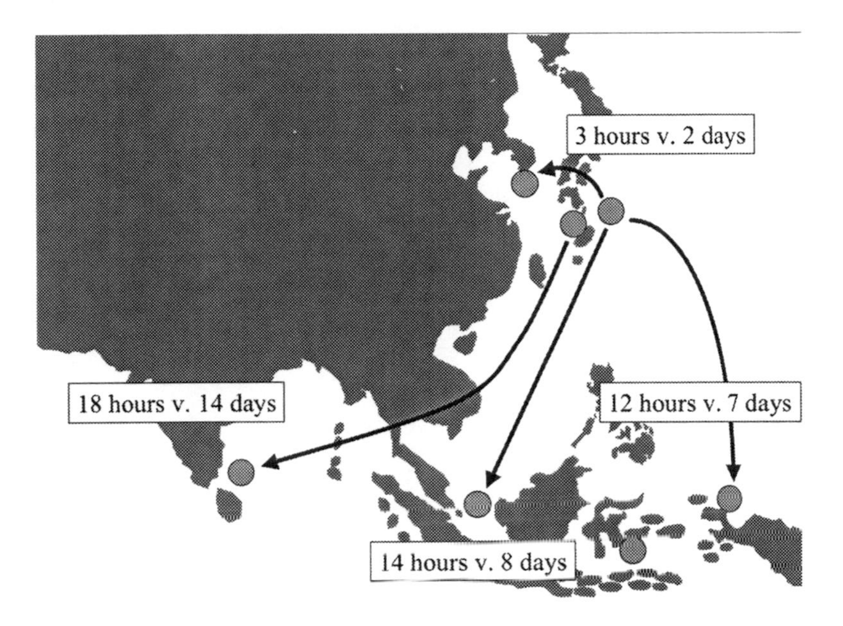

Figure 6.10—Operational Speed Redefined.

As a final point, a fully constituted distributed networked force has an additional benefit over traditional forces. It is well known in network research that one cannot turn a chain into a complex network but a chain can be made from a complex network. In distributed operations, traditional forces cannot be distributed networked forces, but traditional force structure can be made from a distributed networked force. Distributed networked operations should provide such unprecedented levels of flexibility that new schemes of maneuver will be possible, inspiring a new wave of innovations to Operational Art.

[1] This vignette was first presented in preparation for a war game series at the CNO Strategic Studies Group in 1998. See LCDR Jeffrey R. Cares, "Adaptive Forces," Presentation to the Complexity Working Group, Newport, RI, 4 Mar 1999.

7

Realizing Distributed Military Operations

Go is an ancient Chinese game of strategy in which players try to capture each other's playing pieces while occupying territory on a 19 X 19 square game board. *Go* has been used by many cultures as both a metaphor for warfare and for training students in strategy. Researchers in Newport, Rhode Island, recently took the standard *Go* board grid that has 361 nodes and 684 links and rewired it into a complex network with roughly the statistics in Table 5.1. Figure 7.1 is the Network-Centric *Go*™ game network. Players immediately realized that the competition was transformed. First, the environment was instantly converted from simple (a grid, or lattice of degree 4) to complex. The new, networked *Go* board could not be navigated without an adjacency matrix, a degree distribution and other network statistics near at hand. Second, they soon discovered the need for additional tools to understand the competition, and began to invent new ways of measuring their progress in the game.

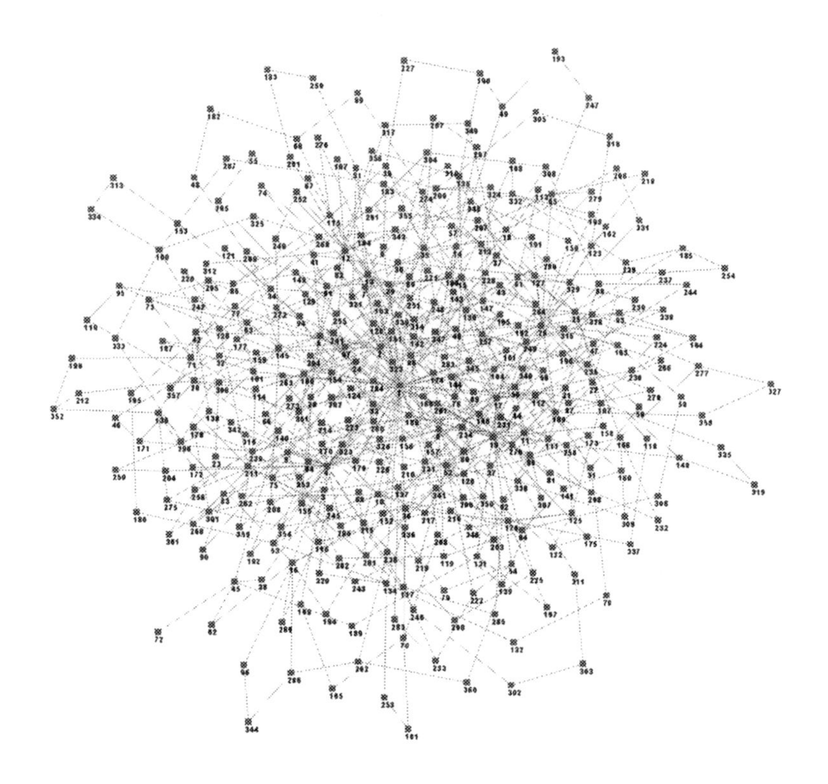

Figure 7.1—Network-Centric Go™

Third, although the basic rules of play remained the same, the players' stratagems (rule sets) changed play dramatically. Recognizing that higher degree nodes were much more valuable than low-degree nodes, players adopted a decreasing-degree strategy: they occupied nodes in decreasing degree sequence. Players initially thought this simple rule would always dominate, but continued play identified other stratagems to counter this first-mover advantage. The decreasing degree rule actually resulted in a classic *Go* strategy: to win by building large contiguous masses of pieces. One winning strategy that emerged in Network-Centric *Go*™, however, was to occupy smaller, non-contiguous clusters of nodes. This allowed players to keep options open until much later in the competition (thereby obscuring potential playing sequences) yet rapidly bring their node clusters together in a large mass by using punctuated growth.[1]

There is a sharp difference between the researchers' experience with this new, complex rendition of a traditional, simple game and the defense community's experience with transformation to the Information Age. After playing the new game for a while, players began to recognize how the new environment and distribution of physical elements transformed the competition. Change in the defense community, however, has been glacial: although the initial concepts for Network Centric Warfare were presented in the mid-1990s, we are well into the mid-2000s and a convincing realization of the power of these ideas is still elusive.

Part of the reason that the full potential of Network Centric Warfare has not yet been realized is that the military community often confuses IT-enhanced, rarefied Industrial Age processes and distributed networked systems that are truly transformed for the Information Age. For distributed networked operations to be more successful, the defense community must correctly evaluate, identify and develop Information Age systems. This chapter explores this issue with respect to the three questions about distributed networked systems[2] that were asked in the introduction:

- What are the defining characteristics of a distributed networked force?
- What should a distributed networked force be capable of doing?
- How should distributed networked systems be developed to exploit their full potential?

DEFINING CHARACTERISTICS

Table 5.1 contains a set of characteristics of a distributed, adaptive system. These quantitative measures constitute *defining characteristics* because they specify the range of values that a distributed networked system should have. Together they constitute a rough blueprint for the structure and dynamics that must be present for effective distributed networked operations. The defining characteristics are:

Number of nodes, N. The networked effects that contribute to adaptation are unlikely to be realized in a distributed system of fewer than 50 nodes.

Link to node ratio, l/N. Distributed networked forces should have about two links per node. Research indicates that even four or five links per node can paralyze a network with too much feedback or feed-forward.

Degree distribution. Distributed networked forces should have a skew distribution of links. This will allow for smooth reconfiguration of network hubs.

Size, connectivity of the largest hubs. The largest hub in distributed networked operations should contain fewer than 100 links and the system should be engineered for survivability by isolating the largest hubs from each other.

Characteristic Path Length. A network with 2:1 link to node ratio and a skew distribution of links will have very short characteristic path lengths. This measurement of distance in distributed networked operations should remain small and grow by no more than the logarithm of the number of nodes interacting during an operation.

Clustering Coefficient. The overall clustering coefficient of a distributed networked force should be between 0.1 and 0.25, meaning that on average about 10- to 25-per cent of a node's neighbors will be neighbors of each other. A skew distribution of clustering coefficients defines the structure of adaptive hierarchy in a distributed system.

Betweenness. A distributed networked system should have a skewed distribution of betweenness to control cascades of pathological behaviors and to prevent bottlenecks.

Path horizon. Distributed networked forces exhibit good self-synchronizing behavior when the path horizon is the logarithm of the number of nodes in the system.

Neutrality Rating. Neutrality is the additional structure in a distributed system over and above the minimum for required for connectivity. A distributed networked system should have a neutrality rating of between 0.8 and 1.2.

Coefficient of networked effects (CNE). The coefficient of networked effects measures the amount of cyclic behavior per node. A distributed networked system should have a CNE between .1 and .25, meaning that about 10- to 25-per cent of the system's connectivity comprises feedback or feed forward loops.

None of these characteristics by itself indicates that a system is a distributed, adaptive system. These measures must be evaluated collectively, much like a doctor examines a patient's symptoms holistically: only in extreme cases will blood pressure or body temperature alone tell a doctor what she needs to know about the health of the patient. Alternatively, there may be times when a symptom is outside the norm yet not be indicative of ill health, such as a high heart rate after exertion. Just because one of the network statistics is in the recommended range does not mean that a network is adaptive. Similarly, a distributed networked system could exhibit some characteristics outside the range of those in Table 5.1 yet still produce networked effects in combat. Moreover, the values as presented are derived from observing good performance in distributed adaptive systems in non-military domains. It may well be that additional research will modify or refine these initial thumb rules into values that are more specific to military distributed networked operations.

All the characteristics list above, however, should conspire to create the behaviors indicative of fully adaptive distributed networked operations: indirect compounding feed-forward and feedback mechanisms that result in a rapid accumulation of network effects. These tipping points can emerge, disappear and then reappear in different locations throughout the network based on evolving environmental conditions. A major source of this adaptability is latent link and node structure. This additional structure is also the mechanism for another important behavior indicative of a distributed networked system: decreased susceptibility to loss of network effects due to link or node removal.

OPERATIONAL CAPABILITIES

Chapter 5 described a model for understanding how the elements of a distributed networked force are connected into dynamic collectives. Although the defense community is on the wrong side of the learning curve with respect to the engineering of a distributed networked system, there are some rudimentary design principles the community can follow that will improve the chances that their efforts will result in a true distributed networked force.[3] Many in the defense community, however, treat such a list as an input to the design process. Each of these principles should be seen as a continuum to be explored by the design process, not a pre-determined point in design space. These principles therefore represent outputs of a design process—a list of *operational capabilities*. An initial list includes:

Recombination. The ability to aggregate, distribute or interchange physical, informational or logical elements and connections. This operational capability provides a distributed force that adaptively evolves to a number of nodes, connection topology and multiscale performance that provide for advantageous adaptive behaviors.

Dispersion. Avoiding spatial, informational, or logical centers of gravity while confounding adversary command, control, and scouting resources. This operational capability provides one of the most significant advantages of distributed networked operations: the ability to keep assets and their interactions obscure and protected by dispersal until they are rapidly connected for application by punctuated growth.

Mobility. Sufficient speed for rapid relocation of elements or element collectives. Mobility provides a repertoire of maneuvers for a range of operational situations, distribution options or recombinations.

Stealth. Distributed networked operations can suggest ways in which smaller elements can improve collective performance yet reduce observability.

Proximity. Most objects in contemporary military systems have such high intrinsic value that direct proximity to a threat incurs great risk to mission success. Distributed networked operations can provide a level of proximity dictated by appropriate connection strengths, collective behaviors and scale considerations. Greater numbers of smaller, stealthier objects, which manifest more of their value in informational and logical contributions than physical combat power complement this operational capability.

Flexibility. Reliable, fluid system substructures with a wide range of interoperability options. Production of this capability would enable a force to adapt to radical competitive behaviors or extreme changes in environmental conditions. A key aspect of this design principle that sets it apart from recombination is the robustness of the force's information technology architectures and information management schemes.

Persistence. Ability of forces to operate without the disruption inherent in traditional logistics structures. This operational capability requires *reduced* constraints on logistics and a more adaptive flow of goods and service (a notion counter to the Joint Vision Concept of Focused Logistics).[4]

DEVELOPING DISTRIBUTED NETWORKED SYSTEMS

Existing DoD systems are engineered from the top, down. At the heart of this approach is the process by which the performance of a military system is prescribed and then is built to that prescription. For major defense systems, the process begins with justification for the system. An Operational Requirements Document (ORD) is drafted to describe the operational capabilities of the system based on the projected operational environment and an analysis of missions that the system is expected to perform, called the Mission Needs Statement (MNS). The ORD is decomposed into functional components and each component is assigned a Key Performance Parameters (KPPs). Defense acquisition officials then contract with private industry to produce the system and evaluate contractor performance based on adherence to these prescribed standards and expectations.[5]

There are three main reasons why this process is problematic for the development of distributed networked forces. First, the defense community is on the wrong side of the learning curve for the science, technology and Operational Art of distributed networked operations. Although much has been written about these systems (including a great deal that suggests they already exist), no one has yet developed a distributed networked force that meets the full promise of the NCW literature. As a result, the defense community is ill informed about their use and development. The second reason is more problematic. Recall from Chapter 5 that distributed networked operations are complex in the sense that there are an extraordinarily large number of different configurations that can be achieved in a distributed networked force. Recall from Chapter 2 that a complex environment is one in which all trajectories to future states cannot be known in advance because of strong dependencies between states. Since distributed networked operations are conducted by a complex system in a complex environment, a traditional MNS is almost guaranteed to misrepresent the missions the system is expected to accomplish, the ORD cannot possibly detail operational capabilities and the KPPs can at best only approach the kind of rough approximations listed in Table 5.1. The third reason is that networks evolve from clusters to the whole—from the bottom, up rather than from the top, down—by the process of punctuated growth described in Chapter 5.

These three points highlight the need for a new method for developing distributed networked systems. The following procedure is proposed as an alternative to the existing DoD research, development and acquisition system:

Step 1: Environmental Assessment. Analysts should review the types of environments in which military forces will be required to operate. Initially, this should be accomplished at the finest scale possible for three reasons. First, coarser-scale descriptions appear simple and might mistakenly suggest a complex force is not required. Second, a distributed networked force will be built "from the cluster, up." Analysis of fine scale environments dictates the requirements of smaller clusters better than analysis of coarse scale environments. Third, since we are largely ignorant about the engineering of these systems, we should start with the most manageable, yet complex, sub-systems we can devise. These small complex systems will be more successfully tested in fine-scale environments.

Step 2: Mission Assessment. Engineers need to know what objectives a distributed networked force should accomplish, so a mission assessment must be conducted. As just discussed, however, a traditional Mission Needs Statement is not useful in distributed networked systems engineering. A new kind of mission assessment for distributed networked operations would begin with a small subset of likely missions in a fine scale environment. Justification for this approach mirrors that in the previous paragraph: coarse scale missions suggest simple forces, fine-scale mission assessment better supports development of small clusters capability and producing a system to satisfy fine-scale missions will decrease engineering complexity in the early stages of development. Of course, missions can be re-assessed at later stages in system development to include more fine-scale missions, more fine-scale environments or to combine more than one cluster of fine-scale missions in to a coarser-scale mission for a coarser-scale environment.

Step 3: Experiment "In Silico." In the top-down acquisition procedures, defense industry engineers are typically focused on the physical elements of a system, such as the vehicles, sensors, weapons or IT hardware. The integration of the parts into a whole is predetermined and assumes the parts are already engineered for seamless integration. As a result, improving the performance of the collective is equivalent to improving the performance of the individual elements. In other words, there is no feedback from the design of the collective into the design of the parts. A properly engineered distributed networked system should be developed by starting with a set of immature physical objects and iteratively comparing individual improvements with collective improvements. This will likely mean trading off between individual functional characteristics (such as detection range and operational speed) and collective characteristics (such as numbers of elements and rule sets).

The next step in engineering a distributed, adaptive system, therefore, is to perform this trade-off analysis by conducting simulation experiments. A distributed networked system is a constellation of many elements collectively performing a mission in some environment. Since a computer simulation is a digital exercise, element functions and potential rule sets can be modified in a digital representation more readily than physical characteristics can be modified on a shop bench. For distributed networked operations, this best performed in a low fidelity simulation, such as an *agent-based simulation*, which allows for a large number of very simple elements to interact and, though successive iterations, for the elements and their interactions to be made more complex.[6]

For a digital design exploration to be successful given the high dimensionality of distributed networked operations, the simulated representations must cover a broad range of input values. The selection of the range of the values, however, can be just as important as the functional representations themselves. If too narrow a range is selected, then only a small part of the solution space will be searched. If too broad a range is selected, then unrealistic inputs may contaminate the output. Moreover, if the interval between values for a particular range of input is too large, some important combinations of input values might not be examined. Finally, if the interval between values for a particular range of input is too small, then the solution space can get far too large to be practically evaluated with anything but the most powerful computers.

This last point bears a deeper discussion. If one were to look at seven different variables over a range of ten values for each variable, then the total number of points in the solution space would be 10^7 = 10,000,000 different data points. Since the experiment should necessarily include randomness in its processes, then each data point must be explored with a statistically significant number of runs, each with a different random seed. If the number of required runs for each point is 100, 1000, or 100,000 (depending on degrees of freedom and the character of stochasticity), then the total number of runs, which is equal to the number of data points multiplied by the number of runs per data point, can start to grow by many orders of magnitude. The type of high dimensionality displayed in Figure 5.5 is a fundamental characteristic of distributed networked systems that will translate into extremely large batches of simulation runs. Since agent-based simulations have lower fidelity than other simulation methods their much shorter run-times are most appropriate for such large batches of simulation runs.

Step 4: Experiment with "Physical Abstracts." Successful *in silico* experimentation should produce first approximations of the number of elements required in the system, the basic functions that these elements should perform and the initial topology of the collective. These approximations should be immediately embodied in a rough prototype physical system. This system, an oxymoronic *physical abstract*, can be used to better understand first order engineering challenges in the system and to explore any physical domain issues that cannot be resolved by simulation. A physical abstract is a set of platforms brought together not to test platforms but to test a collective of platforms. Physical abstracts should be kept primitive so they can be easily and quickly modified based on early lessons learned from experimentation. This also guards against conflating improvements to physical elements alone with improving the behavior of the collective.

Step 5: Feedback Between Experiments. The results of *in silico* and physical abstract experiments should iteratively feed into each other to determine the small number of elemental collectives that show the most promise over the widest range of operational environments. This step should result in a more refined approximation of the element collectives, to include more detail on the number of elements required in the system, a more advanced assessment of the functions that these elements should perform and a more specific topology of the collective to include not just the topological structure, but the identification of networked effects and a description of the network's long-term evolution. The focus of this step should be to co-evolve rule sets for adaptive collective behavior with individual design characteristics.

Step 6: High Fidelity Simulation. Increasingly refined experimentation should result engineering-level detail sufficient for a high fidelity simulation of distributed networked operations. This step should include very specific representation of the system, the environment as well as competition against an adversary system. This step is analogous to the Analysis of Alternatives (an analytical "fly-off" of different system proposals or force structures) required by traditional acquisition process. Based on the results of this step, the candidate collectives either return to Step 5 for more experimentation or proceed to production and testing.

Step 7: Production and Testing. The collective is now ready for traditional prototyping, testing and production. In this step a prototype system is built and tested. This version of the system is a true prototype, intended to help engineers proceed with engineering development (as contrasted with prototypes in many mature industrial processes that are in fact advanced, next-generation products). Engineers will incorporate what they learn from building the prototype into production run design as well as feed what they learn in testing the prototype back into the high fidelity simulation to further refine production run versions. These feedforward and feedback loops should remain for the duration of the program.

Step 8: Continuous Operational Assessment. Every distributed networked force should be fielded with the direct means for operators to feed their lessons learned, operational assessment and request for new functions, missions and environmental coverage back into experimentation and engineering.

These steps should proceed sequentially during the very first attempts by the defense community to build a distributed networked system. As analysts, engineers and operators gain experience and confidence with the analysis, engineering and products of this process, the feedback and feed-forward loops should become more dense, creating staggered development sequences or shortcuts for rapid-prototyping. As the engineering of distributed networked forces advances to a more mature engineering field, it would seem quite natural for analysts to feed the results of Step 7, for example, directly into Step 2 or Step 8 into Step 4. This process for engineering distributed networked systems will become adaptive, using feedback and shortcuts to create preferential attachment, punctuated growth and networked effects within the production system itself. Over the long term, the engineering process would resemble not a production line, but a production network, perhaps with a topology approximated by the values in Table 5.1.

Many familiar with software development might recognize the same philosophy in spiral development, which includes increasing sophistication with each release, frequent feedback from users and re-use of existing products wherever possible. What sets distributed networked systems development apart from existing spiral development efforts, however, are two factors. The first is that successful development of distributed networked systems will create an engineering discipline where none currently exists. The second is that distributed networked systems development will rely on non-traditional technical competencies and cross-disciplinary approaches.

Distributed Networked Engineering

A very good analogy for the maturation process for engineering distributed networked forces is the process by which Aeronautical Engineering grew to become a technical field in its own right. Before man took to the air in fixed-wing flight, the small number of pre-aviation pioneers came from many different scientific backgrounds. Some, like the Wright brothers, were not scientists at all, but bicycle shop owners who were also self-taught practitioners of a nascent science of flight. Some of the basic physics, thermodynamics and fluid dynamics were known, but the application of these scientific principles had to be developed by practical adventurers—one could certainly not go to the universities of the day to study aircraft design.

The internal combustion engine, wind tunnels and rudimentary aircraft fabrication techniques were already in place in 1903. Once the Wrights succeeded, a flurry of engineering activity followed. Soon, a great diversity of efforts exploded on the aviation community, as the incipient aircraft industry produced, for example, monoplanes, biplanes and triplanes; engines that pulled and engines that pushed; inline engines and radial engines; single engine aircraft and multi-engine aircraft; aircraft with pilots lying prone, sitting up, sitting side-by-side or sitting in line; etc. Eventually, engineers developed a sense for what configurations were most airworthy, which were most economical for production and which performed best in different environments. Dominant designs emerged from this growing body of practical knowledge.[7] Over time, additional technologies were added to aircraft designs, such as radios and electricity, hydraulics for flight control and weapons systems. A growing aviation industry also found new uses for aircraft, including air mail, military scouting, long range bombing, passenger travel—even entertainment in the form of barnstorming and air shows. Each of these new uses brought their own engineering challenges. It was not until many decades after the first flight, however, that engineers had learned enough about how to build airplanes that the study of airplane design, Aeronautical Engineering, became institutionalized as a formal engineering pursuit.

Since institutionalized engineering education is much more advanced than in the early part of the last century, one should not expect the field of distributed networked engineering to take decades to mature. A few other expectations, however, are appropriate. First, one should not expect distributed networked engineering to antiseptically arise from academic study of the topic: we must build, learn, and build some more—and what we learn must be informed by use, preferably in market, military or other types of competition. Second, we must expect a diversity of developmental efforts. No one should prematurely declare an engineering method, control algorithm, design specification or product dominant. Third, learning must come from both success and failure. Developers (and government funding sources) must be willing to risk failure during development. Finally, researchers and engineers should start now to institutionalize the study of distributed networked forces, so that the lessons learned by the community—particularly those learned at great cost—can be collected in a way that speeds the development of distributed networked engineering and allows a new community to quickly climb a challenging learning curve.

CROSS-DISCIPLINARY APPROACHES

Many early innovations in aviation engineering were the product of cross-disciplinary activities. Similarly, the early successes in developing distributed networked systems will be the result of a necessity-driven combinations of scientific and technological disciplines that are not typically connected in legacy defense programs. The research contained in this book, in fact, has drawn heavily on such cross-disciplinary research—complex systems research—recently initiated to understand distributed networked behaviors in other human endeavors.

Contrary to many popular claims, there is no overarching "Complexity Theory," but research into complex systems has nonetheless constituted a synthesis movement in academia, coupled with advances in computing power, that encourages exploration of phenomena previously impenetrable by classical analytic methods. This work is changing the understanding of mature disciplines and giving rise to new, cross-disciplinary research pursuits. Some in the popular science press have heralded complex systems research as the "New Sciences," but many of the new ideas that are emerging from this "new" science trace their origins to classic science. Just as the operational employment of distributed networked systems should carry some (but not all) of the collective experience from building and operating legacy systems into engineering of new systems, the science behind the engineering of distributed networked systems should build on the long pedigree of classical science.

Complex systems research will therefore inform the engineering of distributed networked systems. An engineering approach that applies this cross-disciplinary research will rely on the substantial scientific literature in such topics as economics, game theory, graph theory, information theory, algorithmic information content, multi-scale representations, genetics, biochemistry, mathematical biology, evolutionary biology, paleo-biology, cellular automata, agent-based modeling, epidemiology, evolutionary computation, sociology and anthropology. Of course, a new generation of scientists and engineers who understand the military applications of these disciplines must be educated, trained and promoted. The technological challenges facing the development of distributed networked systems are substantial indeed. Cultural change, however, is often much more difficult than technological change, so the challenges in establishing a new generation of Distributed Networked Systems Engineers may prove to be even tougher. Without both the engineering and cultural transformations, however, distributed networked operations will continue to fall far short of their full potential.

[1] Network-Centric Go™ Board ©2002-2004, Jeffrey R. Cares and Valdis Krebs. The game and lessons for networked competition are outlined in a forth-coming article in The Information Age Warfare Quarterly, http://www.iawq.com.

[2] These three questions were the inspiration for the research that coined the term "distributed, networked forces," Jeffrey R. Cares, Raymond J. Christian, Robert C. Manke, Fundamentals of Distributed, Networked Military Forces and the Engineering of Distributed Systems, NUWC-NPT Technical Report 11,366, 9 May 2002, NUWC Division Newport.

[3] Cares, et al., 20-22. The following list is a verbatim transcription from that publication.

[4] See Cares, Jeffrey R. and CAPT Linda Lewandowski, USN, Sense and Respond Logistics: The Logic of Demand Networks, undated, unpublished US Government white paper, 2002, for a discussion of distributed networked logistics systems.

[5] These and other acquisition topics can be reviewed at the Defense Acquisition Resource Center website, http://akss.dau.mil/darc/darc.html, accessed 16 Jan 2005.

[6] See http://www.dnosim.com for software and support for simulation of distributed networked operations.

[7] This idea of technological evolution as diversity leading to dominant technical designs is attributed to Dr. Stuart Kauffman, who uses the aviation example in his public presentations to describe fitness landscapes and methods of natural and artificial evolution.

Appendix I: Commercial Applications

For more than a decade, many in the defense community have lamented that they have lagged private industry in developing innovative technologies, Information Age value propositions and new process models. This lag, however, is not necessarily a permanent condition; in fact for most of the post-World War II era the military enjoyed a substantial technical lead over the private sector in many industries. The insights, innovations and break-through research that will result from success in developing distributed networked systems will once again give the military a leading role in developing new technologies and processes for the Information Age. This appendix briefly discusses ten areas where there will be commercial spin-offs from the science, technology and operational concepts behind distributed networked military systems.

LOGISTICS

Long-standing practices in logistics and supply chain management work best in environments where there are high levels of stability and predictability. In other words, they are not designed to cope with the quickly evolving and adaptively *ad hoc* behaviors envisioned in Information Age business models. Most concepts proposed for Information Age logistics suggest applying Information Technology (IT) to marginally improve existing processes and do not address the need (particularly at the local level) for Information Age logistics processes that are as robust and dynamic as the business models they would serve. Early research into distributed, adaptive systems introduced the Sense and

Respond Logistics concept. This is a theory of logistics management for supporting distributed networked forces in challenging environments that values flexibility, adaptation and learning over predictability, precision and optimization.[1] This concept and the research supporting it have direct application to commercial supply network management.

TRANSPORTATION

One of the early results from research into Sense and Respond Logistics was the immediate need for more transportation assets in the Information Age battlespace. This complements other trends toward developing future forces with more numerous, smaller vehicles to support Information Age operations. Chapter 1 briefly discussed some of these concepts and Chapter 7 provided an operational vignette of how a transformation in transportation will contribute to distributed networked systems. Successful development of distributed networked operations will inform new business models that rely on new vehicular fleets for such industries as overnight shipping, short sea freight, personal transportation and commercial travel.

MANAGEMENT THEORY

Throughout the Industrial Age, the military community depended on a model developed by Frederick Lanchester for force structure assessments, acquisition decisions, manpower policies and campaign analyses. Similarly, classic Industrial Age management theory relied on the philosophies of another Frederick of the same vintage, Frederick Taylor.[2] For the last decade, "New Economy" business gurus have been searching for Information Age business models to replace this Industrial Age classic. The military's success in managing the development and operations of distributed networked forces will be a substantial contribution to the Information Age business knowledge base, providing practical advice to business leaders and academics as they continue to wrestle with this difficult topic.

PRODUCTION

Chapter 4 considered the problem of *makespan* in production lines. Just as many Industrial Age military processes are inherently linear and inflexible, there are numerous processes in the commercial sector that exhibit such classic production line characteristics. Some of the ideas that inspire development of distributed networked operations are already being used to revolutionize automobile fabrication, cosmetic production and pharmaceutical development. Successful production methods discovered through engineering and fielding distributed networked military systems will be a significant addition to Information Age production engineering.

BUSINESS MODEL INNOVATION

Chapter 5 contained an Information Age Combat Model that was developed to help formalize the exploration and identification of the sources of value in Information Age military competition. As the Information Age progresses, other industries must develop similar models to identify the important feedback mechanisms, networked effects and value propositions that contribute to success in their respective markets. The Information Age Combat Model, combined with successful development and employment of distributed networked forces, can act as an important source of insight for such activities.

Customer Relationship Management

The Information Age has brought new dimensions to the manufacture and delivery of consumer goods. It is now possible to track consumer related information such as individual servicing requirements, customer profitability and brand loyalty, creating new feedback loops between customers and the products they buy. These and other related activities are finding a larger role in new business models under the rubric of Customer Relationship Management (CRM). CRM efforts, however, can be extremely inflexible when they executed by large call centers, centralized online help-desks or consolidated customer support. A decentralized, adaptive CRM capability could be developed as a spin-off from military success in managing the human interactions and interfaces with distributed networked systems.

Product Development

Defense product development cycles take years, sometimes decades, to mature from the drawing board to the field. In addition, these cycles are typically very high cost. Distributed networked engineering is likely to reduce new product development cycle time and costs substantially, particularly as modularized products are developed that adaptively respond to emerging threats and opportunities. As the military gains experience in successfully transforming from Industrial Age procurement to Information Age adaptive acquisition, this knowledge can be used to redesign the product development cycles of industries with very long capital time constants.

Human Capital

One of the most difficult tasks in management is the identification of human potential and efficient assignment of human capital. As industries migrate from production line manufacturing to more cognition-based service products, the ability to identify the topology of clusters of cognitive skills in a work force and to adaptively reassign this talent as quickly as market demand shifts will be highly prized. Experience with the substantial cognitive activities required to operate a distributed networked force will contribute significantly to business applications of adaptive human capital strategies.

Marketing

Industrial Age economies of scale required that a firm had to either accurately predict which few products the mass of consumers would buy or expend extraordinary advertising and marketing efforts to create demand. In the Information Age, another byproduct of new feedback loops between customers and the products they buy is the potential for mass customization to replace mass production. Most mass customization efforts rely on IT-enhanced inputs to the production process. Simply reacting to customer preferences, however, is inferior to an ability to adaptively anticipate customer tastes. The military's experience in adaptively responding to competitive requirements could form the basis for a new generation of adaptive marketing concepts with the potential revolutionize such techniques as mass customization.

Value Chain Analysis

Each of the preceding nine commercial applications of distributed, adaptive systems science, technology and operational concepts is an input to the business process. A collection of these is known as a *value chain.* Value chain analysis is an important business management activity that helps decision makers determine investment strategies, management methods and trade-off decisions. The military's understanding of the value proposition behind distributed networked forces will shed new light on how all of a firm's inputs to production contribute to long-term profitability in the Information Age.

[1] Cares, Jeffrey R. and CAPT Linda Lewandowski, USN, Sense and Respond Logistics: The Logic of Demand Networks, undated, unpublished US Government white paper, 2002, for a discussion of this concept.
[2] See http://www.fordham.edu/halsall/mod/1911taylor.html for one of Taylor's essays on "Principles of Scientific Management."

Appendix II: Complex Network Primer

The research supporting this book included an extensive examination of network flows and graphs, including very recent research into complex networks, much of which is still developing. This appendix is a primer that presents the findings of this research that is most relevant to understanding distributed networked operations.

NETWORK THEORY

What is commonly called a *network* is more technically described as a *graph*. A graph is a simple collection of *links* and *nodes*. When values are assigned to the links and nodes, a system with its own logic is created. This system is more properly called a *network*. Networks are typically used to mathematically model flows, analyze network circulation or evaluate costs in a dynamic, distributed system. For the purposes of this primer the values can be removed, greatly simplifying the discussion without loss of validity. Properties that characterize the performance of networks include:

- Link/node Ratio: Compares the link densities of different size networks.
- Characteristic Path Length (CPL): The median (middle value of ranked values) of the average distance from each node to every other node in a network.[1] A short CPL means that commodities proliferate through a network without passing through a high number of nodes.[2]

- Diffusion Rate: Describes the rate at which commodities proliferate throughout a network.
- Clustering: A measure of local cohesion in a network. The clustering coefficient, γ, is the ratio of the number of actual links between neighbors to the number of possible links between neighbors. In a social context, this would be a measure of how many of one's friends are also friends of each other. Highly clustered networks tend to have pockets of connectivity, which can increase the connectivity and redundancy of the whole network.[3]
- Scale: A measure of the distribution of links among nodes in a network. If the distribution is uniformly or normally distributed, then the network is said to have a definite scale. If the distribution belongs to the family of skewed distributions (similar to the distribution of wealth in some societies), then the network is said to be *scale free*. A scale free network has links distributed according to a Power Law, where the probability that a node has exactly k links is $P(k) \sim k-^{b}$, where b is called the *degree exponent*.[4]

The following examples show how the characteristics describe distinctly different behaviors in different networks.

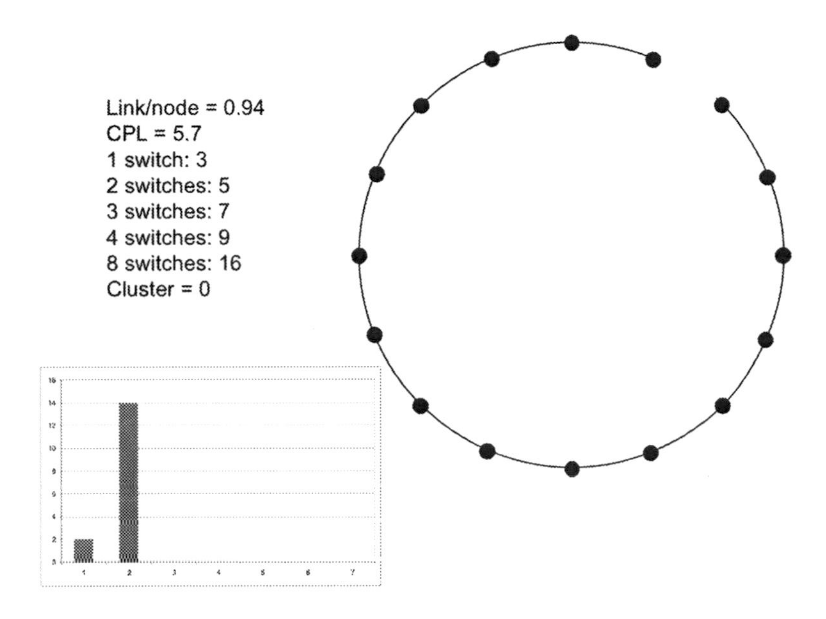

Figure II.1—Minimally Connected Network

Minimally Connected Network. A connected network is one in which every node, *n*, is attached to the network by at least one link. A minimally connected network, also known as a chain, is one in which the nodes are all connected with the minimum number of links possible, i.e., N—1 links. Figure II.1 shows a minimally connected network with 16 nodes and 15 links. In general, a minimally connected network contains:

$$\sum_{i}^{n-1} i$$

different sub-networks, that is, each new node adds *N*-1 sub-networks to the cumulative total of sub-networks (a third node adds two sub-networks to an existing two-node sub-network raising the new cumulative total to three sub-networks; the fourth node brings the cumulative total to six, the fifth brings the cumulative total to ten, etc.). The number of subnets in Figure II.1 is 120. Minimally connected networks have fewer links and fewer subnets than any other connected network and are therefore the cheapest and simplest connected networks, but they have less redundancy and commodities take much longer to proliferate among the nodes. Note, for example, the relatively high CPL, which is represented by the entries in Figure II.1 listing the average number of nodes reachable from each node in N "switches" (which also represents the diffusion rate). Even after 4 switches, each node on average can reach only 9 nodes (including itself). Also note the graph in the lower left, which portrays the number of links attached to a node (the *degree* of the node, horizontal axis) and the number of nodes in each category (vertical axis). This graph defines the *scale*, or *degree distribution* of the network, which in this case is very close to two, because the majority of nodes are connected with only two links. Note also that the clustering coefficient is zero, which indicates that there is very little local network structure in this type of network.

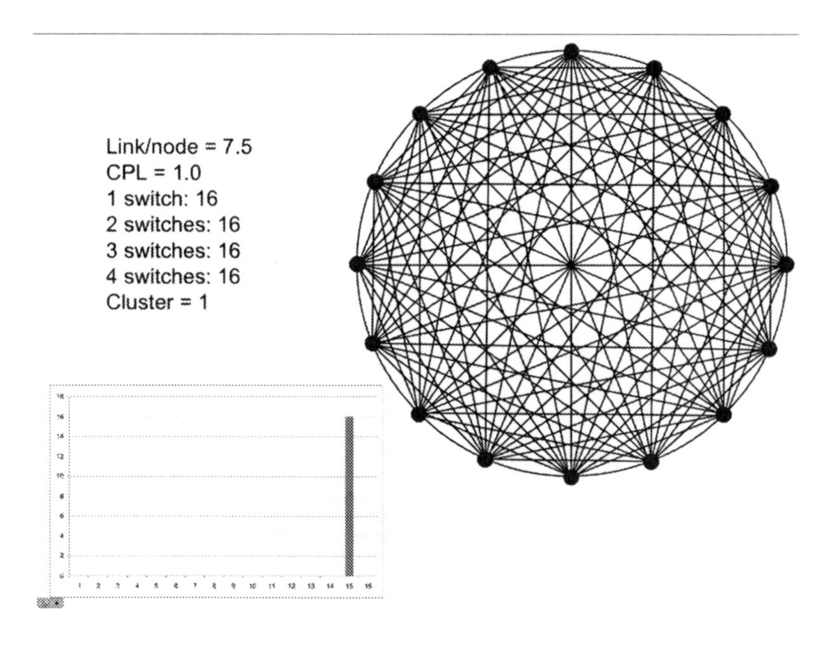

Figure II.2—Maximally Connected Network

Maximally Connected Network. A maximally connected network is one in which every node is directly connected to every other node by one link with

links, that is, each new node adds N-1 links to the cumulative total of links (the third node adds two links to the previous cumulative total of one link raising the new cumulative total to three links; the fourth node brings the cumulative total of links to six, the fifth brings the cumulative total to ten, etc.). Figure II.2 shows a maximally connected network with 16 nodes and 120 links. A maximally connected network contains N! different sub-networks.[5] The number of subnets in Figure II.2 is over 20 trillion. Maximally connected networks have more links and more subnets than any other type of connected network and are therefore the most expensive and complicated connected networks. They have more redundancy and commodities are proliferated more quickly to the nodes (that is, they have the shortest possible characteristic path length). The fundamental drawback of maximally connected networks is that the number of subnets can easily overwhelm attempts to use them efficiently (that is, each flow calculation for the network in Figure II.2 requires over 20 trillion calculations). The scale is fixed at N—1 = 15, and the network is maximally clustered.

Random Network. Minimally and maximally connected networks represent the extremes of network connectivity. For most warfare network applications, neither of these two extremes is useful. Figure II.3 shows an alternative structure, a randomly connected network.[6] The ratio of links to nodes in this network is 2.0 (that is, there are 32 links, about twice as many as the minimally connected network in Figure II.1 yet only about a quarter of the links in the maximally connected network in Figure II.2). The characteristic path length of this network is about halfway between the minimally connected network and the maximally connected network. The random network therefore, is more redundant and commodities are proliferated more quickly than the minimally connected network yet the number of links and subnets is dramatically lower than the maximally connected network. Two drawbacks arise from the random connection of links and nodes. The first is that the network is *irregular* in the sense that CPL has a large variation from node to node. The second is that the network is irregular in the sense that there is a large variation in the clustering coefficient. Irregular path lengths and clustering can cause great unpredictability in networks. Note that the scale of the network spreads out with a peak at about 3.0. If more nodes were added, a smoother bell-curve (Normal distribution) would emerge (although the peak would move more to the right). This portrays a property of random networks: the links are distributed with a Normal distribution with the network scale defined by the peak of the resulting bell curve.[7]

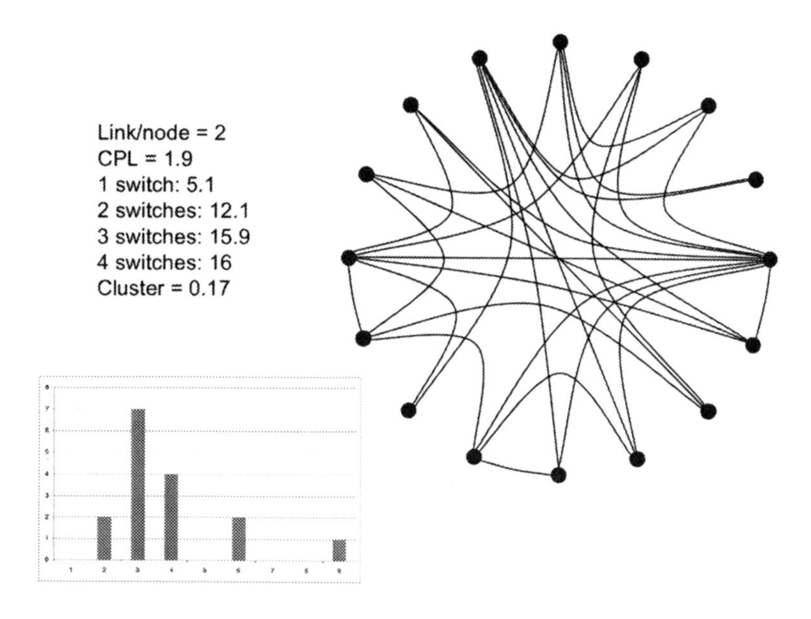

Figure II.3—Random Network

Regular Network. Figure II.4 shows a regular network (otherwise known as a "lattice") that has the same ratio of links to nodes as the irregular random network. Although the clustering of this network is uniform, and therefore more regular than the random network, the characteristic path length increases significantly (although it also becomes more regular). The scale of this network is set at 4.0. Note that the minimally and maximally connected networks are special cases of regular (lattice) networks. Note also the dramatic difference in degree distribution between the regular and random networks, although the number of links and nodes is identical.

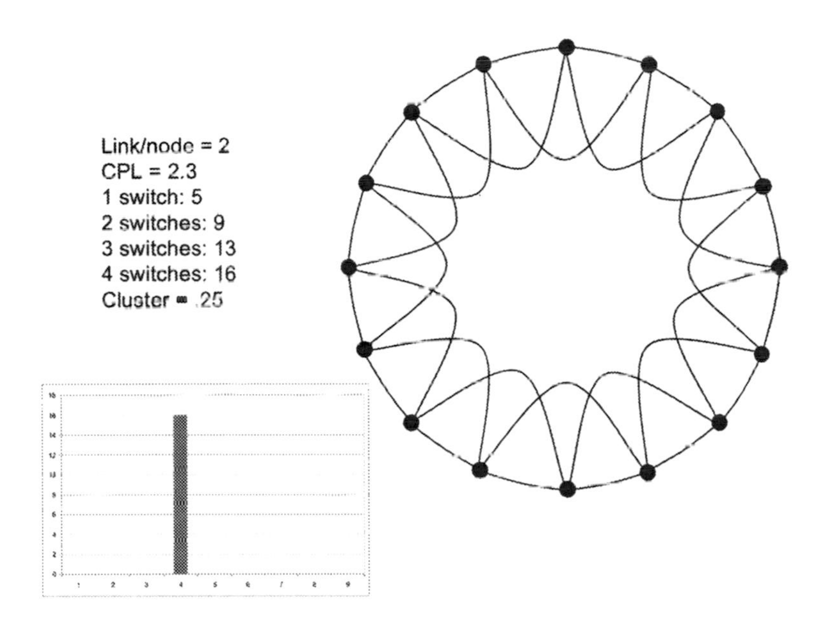

Figure II.4—Regular Network (Lattice)

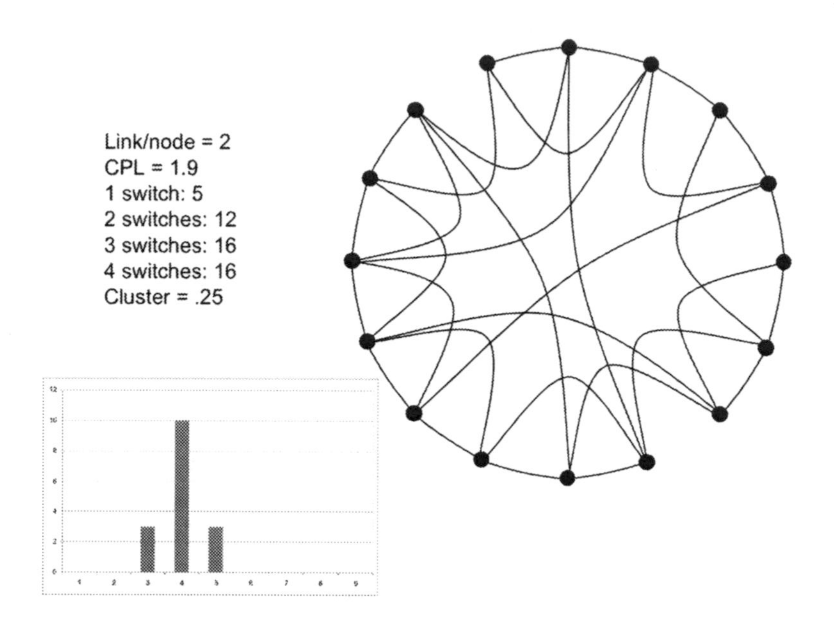

Figure II.5—Small World Network

Small World Network. A minor re-wiring of the regular network can create a *Small World* network that is very regular and has good clustering and short characteristic path length. In a Small World network, remote clustered groups share members with other remote groups so that the average number of links connecting all members remains small (just like handshakes in its cultural counterpart). Figure II.5 shows how the regular network in Figure II.4 can be re-wired to create a Small World network.

Random Network with Growth. For many decades, graph theory research depended on two assumptions that ultimately became obstacles to understanding the network structures found in Information Age processes. These two assumptions were, first, that analysis and theoretical investigation is applied only after all nodes are connected to a network and, second, that links are always added according to a fixed distribution. The network in Figure II.6 shows what happens when network structure is not constrained by the first of these assumptions. This network experiences *growth*, in that new nodes are added to the network as the number of links grows. An obvious result of networks with growth (in this case, with random connections) is that the oldest nodes are most likely to have the highest degree because old nodes have more connection opportunities.[8] The very first node in the network, for example, has N-1 opportunities to connect by the time the Nth node is added. This dynamic, known to economists as a type of *network externality*, has been used to explain "first mover advantage" in the Information Age marketplace. Note that the network is about as clustered as the random network, yet the scale is less defined (this network, in fact, has two distinct scales: degree 1 and degree 2). Also note that although this network has only half the links of the random, lattice and Small World networks, the network still has a fairly good clustering and the CPL grows by no more than about 50-per cent.

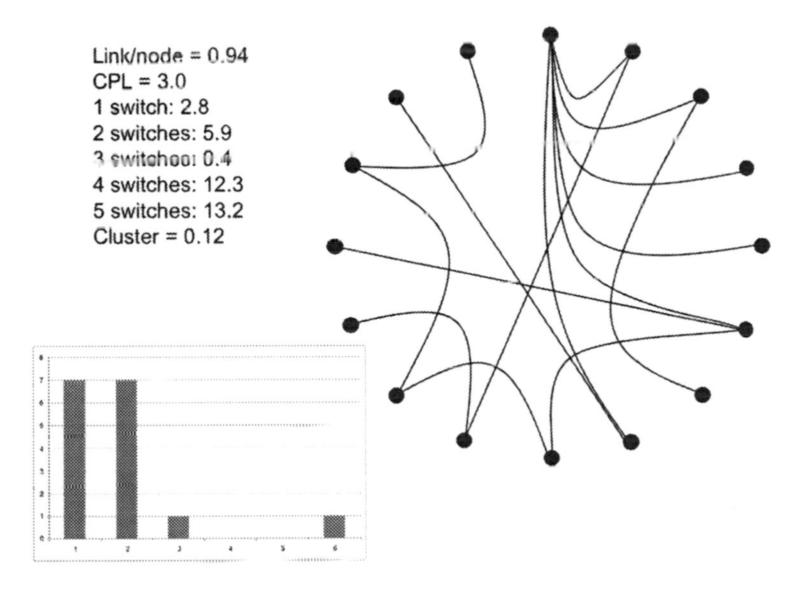

Figure II.6—Random Network with Growth

Scale Free Network with Preferential Attachment. If the constraints of both assumptions are removed so that the network is grown and the connection of nodes is biased, then a class of networks is created that represents many real world networked structures. The network in Figure II.7 was grown by iteratively attaching each new node to a node in the network biased by the number of links each node already possesses. Technically, this was achieved by weighting the probability that a node is selected by the degree of the node. This rich-get-richer scheme is another type of network externality. It is also the type of attachment mechanism that mimics the distribution of routers connections on the internet, the distribution of links to web pages on the world wide web, and a host of other adaptive, dynamic network topologies.[9] The statistics of this network are quite different than the previous examples (the network is not as well clustered as the others and the CPL is almost as long as in the lattice) but it has one beneficial property that marks it as a very adaptive network—it is a *scale free* network. The degree distribution is represented by a skewed curve like the one approximated above the histogram in Figure II.7. One generic form of the equation defined by these curves is the Power Law, but other skewed distributions can represent connections in a scale free network.[10] A scale free distribution of links defined by a skewed distribution has very many nodes with a very small degree, a moderate number with a moderate degree and a very few with a very high degree.

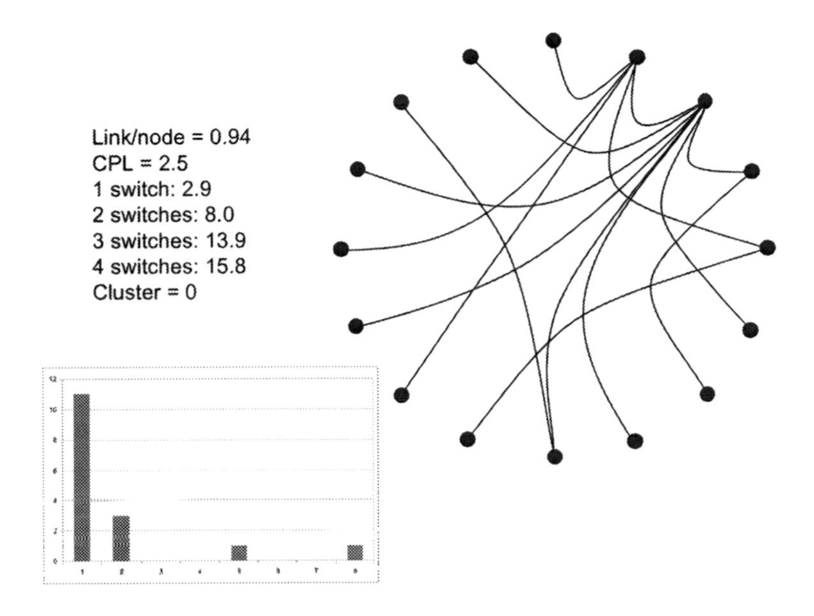

Figure II.7—Growth and Preferential Attachment

Network Type	*l*	*N*	*l/N*	CPL	1 Link	2 Links	3 Links	4 Links	γ
Minimally Connected	15	16	0.94	5.7	3	5	7	9	0
Maximally Connected	120	16	7.5	1.0	16	16	16	16	1
Random	32	16	2	1.9	5.1	12.1	15.9	16	0.17
"Regular"	32	16	2	2.3	5	9	13	16	0.25
Small World	32	16	2	1.9	5	12	16	16	0.25
Random w/ Growth	15	16	0.94	3.0	2.8	5.9	9.4	12.3	0.12
Pref. Attach.	15	16	0.94	2.5	2.9	8.0	13.9	15.8	0

Table II.1—Network Comparison

NETWORK COMPARISON

Table II.1 is a summary of the statistics from the networks described in the previous section. These statistics are the number of links, the number of nodes, the ratio of links to nodes, the CPL, the average number of nodes reached by traversing some number of links (or the number of "switches", listed for 1 to 4 links) and the clustering coefficient, γ.

The minimally connected network has a low ratio of links to nodes, yet the characteristic path length is high. This is because the average number of additional nodes reached for each additional link length traversed increases only by two for each additional link. The CPL is 1.0 in the maximally connected network (that is, every node is directly connected to every other node), yet because of this, deconfliction overhead grows factorially by the number of nodes. The minimally connected network has no clustering and the maximally connected network is maximally clustered.

Random connection of links and nodes with a link-node ratio of 2.0 can connect all nodes in only about 3 switches, CPL is low (1.9) and clustering is also better. Although the random network provides better performance than the minimally connected network and avoids the overhead of a maximally connected network, the network is irregular. One measure of system regularity is the *standard deviation* (probabilistic spread) of measurements within the system. The standard deviation of γ for the random network is 0.15, which means that γ for some nodes may approach 0 (a value of γ similar to a minimally connected network) and γ for some nodes may approach 0.32 (a value of γ similar to regular and small world networks). Creating a regular network from the same number of links and nodes reduces the irregularity in γ but the CPL is much longer.

The Small World network has the same ratio of links to nodes as the random and lattice networks, but retains regularity and clustering while proliferating commodities quickly. The Growth and Preferential Attachment (scale free) network has as many links per node as a minimally connected network, yet has a CPL less than half of a minimally connected network and commodities diffuse much more quickly. Although γ for this sample network is 0, adding just a few more links would start to create clusters around the most well connected hubs. In general, the clustering coefficient for scale free networks is quite low away from hubs and much higher in the vicinity of hubs. This occurs because the hubs are centers of value-creating activity, that is, the motive for preference in preferential attachment. As more links are added, clustering coefficient becomes skewed, with high clustering around the largest hubs, medium clustering near medium hubs and very low clustering away from the hubs.

The preceding analysis demonstrates that the arrangement of links and nodes affects the behavior and performance of a network. Operational requirements often determine this arrangement. Some theories of Information Age warfare refer to "fully-netted" forces; confusing "fully-netted" with maximal connectivity will produce unnecessary cost and complexity. Minimal connectivity, however, will not produce satisfactory network performance and redundancy. Therefore, the connectivity of warfare networks must be at some "sufficient" level. Table II.1 suggests that Small World and Preferential Attachment networks, both in a class called complex networks, are simpler, perform better and require lower overhead than other networks. These networks should be investigated as better models for warfare in the Information Age.

The networks listed here are mathematical abstractions of real-world phenomena. In real-world networks, the operational requirements for which a network is used define how the network will be configured. Moreover, the rationale behind the design is derived from organizational principles and organization theory. The best configuration for a network should therefore be an extension of the purposes and intent implied by the function, roles and behavior of the agents that operate the network, the nature of the tasks required of the networked group and the physical restrictions that may impact allowable connections. Based on the statistics presented in Table II.1, the following comparisons can be made between these networks:

- Minimally Connected Networks are brittle, have long CPLs, poor clustering and definite scale
- Maximally Connected Networks are robust and have the shortest CPLs possible, but they are too clustered (each node is a neighbor of every other node) and have too many links per node. They have definite scale
- Regular Networks are robust but have long CPLs. They are highly clustered and have definite scale
- Random Networks are brittle but have short CPLs. They have low clustering and have definite scale
- Small World Networks are robust, have short CPLs and high clustering and are less scaled
- Random Networks with Growth are less brittle and have short CPLs. They have low clustering and are less scaled
- Networks with Preferential Attachment are robust, with short CPLs and low clustering. They are scale free

DESIRABLE NETWORK PROPERTIES

Most current research on complex networks focuses on discovering the statistical properties of existing complex networks such as the World Wide Web or a sociological data set. One of the aims of this book is to answer the obverse question: if we could choose the type of combat network we should design, what properties should it possess? The following section defines some of the more useful network properties and recommends values for combat networks.

Node and Link Types. A combat network will have many different types of nodes and links.

Flow. Combat networks should capitalize on the existence of cycles and the properties of autocatalysis and neutrality. Such networks will be *directed*, where a node can have links that may be outgoing, incoming, or both.

Number of Nodes. Many of the more important and exploitable networked effects are difficult to achieve unless a network contains at least about 100 nodes. Networked effects become increasingly difficult to achieve as combat networks get smaller.

Number of Links. Although early Network Centric Warfare concepts suggested that each node should be directly linked to every other node for best performance (that is, about N-1 links for every N nodes), most adaptive, complex networks have only about 2N links per N nodes without suffering noticeable degradation in performance. Indeed, having fewer links provides a kind of economy that reduces coordination overhead (as well as the overhead required for protection of links) without adversely affecting performance. Combat networks should therefore have about two links for every node.

Degree Distribution. One way to represent the connection pattern of a network is by the degree distribution, which shows the number of nodes with specific degree. Most adaptive, re-configurable and resilient networks have a skew degree distribution (such as is found in a scale free network). These networks have very many nodes with very few links, a moderate number of nodes with a moderate number of links, and very few nodes with very many links. Skew-degree networks contain powerful hubs that can be adaptively reconfigured. Combat networks should have skew degree distributions.

Maximum Degree. In skew-degree networks, the maximum degree is roughly proportional to square root of the number of nodes. The Figure

II.8 plots maximum degree against number of nodes for a combat network with a skew degree distribution.

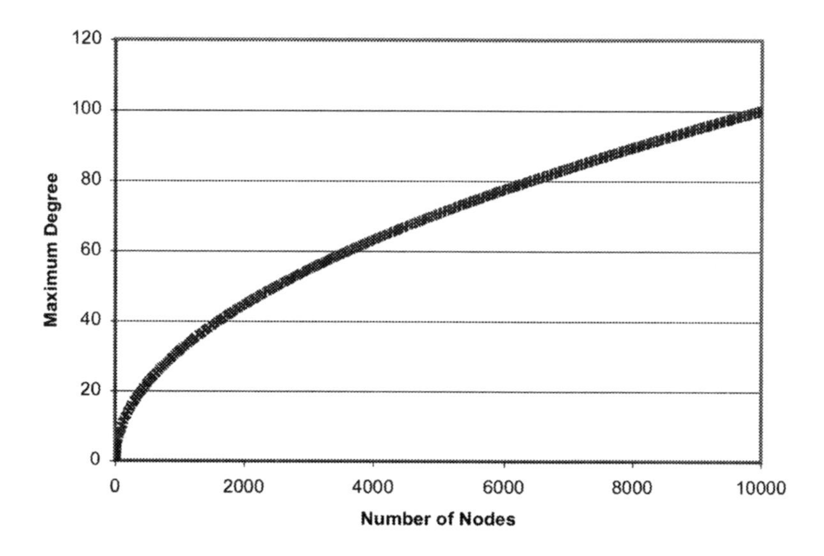

Figure II.8—Maximum Degree in Scale Free Networks

Betweenness. Betweenness is a measure of the number of shortest paths that travel through a node as well as its importance to dynamic behaviors in a complex network. Betweenness can identify the most popular nodes and locate bottlenecks in a network. Research shows that betweenness should have a skewed distribution so that there are not a large number of well-traveled nodes. Also, the nodes of highest betweenness should not be directly connected to each other. This slows the proliferation of pathogens, viruses or cascading damage.[11]

Path Horizon. Path horizon measures the number of nodes on average that a node must interact with for constructive self-synchronization to occur. A path horizon of 1 means that a node must coordinate with all its nearest neighbors for self-synchronization to occur. A path horizon of 2 means that coordination should extend to all the nearest neighbors of a node's nearest neighbors, etc. Recent research suggests that self-synchronization occurs when the path horizon is the logarithm of the number of nodes.[12]

Characteristic Path Length. Figure II.9 lists the CPL of different networks. The legend in the figure refers to lattices of degree 1-5 (e.g., Lat1), random networks of degree 2-5 (e.g., Rnd3), and Small World and Scale Free Networks (SmlWrldSclFree). The CPL of lattices grows on the order of n/4*k*, where n is the number of nodes and *k* is the mean degree. The CPL in random graphs grows proportional to log n/log *k* and in Small World and Scale Free networks the CPL is proportional to log *k* or slower. Combat networks should have characteristic path lengths on the order of log *k* or shorter. This means that for networks as large as 10,000 nodes, one would expect the average distance between nodes to be no more than about 4.[13]

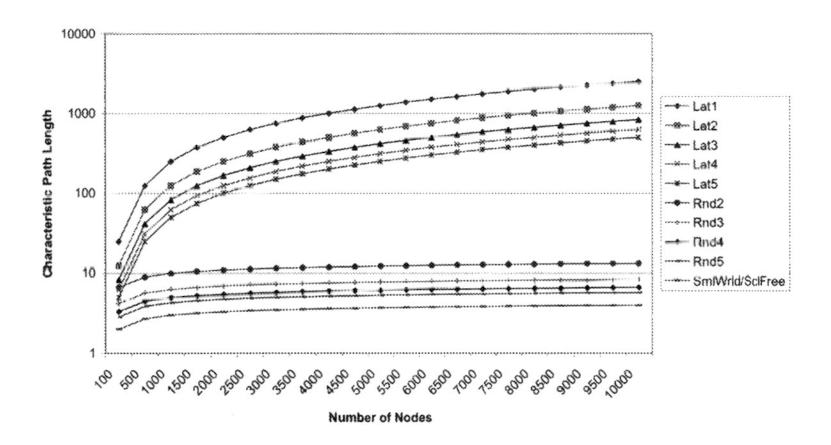

Figure II.9—CPL for Selected Networks

Component Sizes. Since no node should be isolated during military operations, the size of the smallest clusters in a network should be greater than one. Service specific operations in a joint environment or independent special operations would comprise larger sub-components, but to keep the ultimate network configuration obscure to an adversary, the giant component in combat networks should not be fully constituted until a large-scale operational is ready to be executed.

Neutrality. Complex networks require latent, *neutral* structure for adaptation. This allows for the creation, removal and re-emergence of hubs by rearranging only about 5-to 10-per cent of the total number of links in a network. One can always turn a complex network into an optimal (non-adaptive) chain by choosing a particular minimally connected path through the network, but a chain cannot be turned into a complex network without the addition of the adaptive (sub-optimal) neutral structure. The Neutrality Rating is obtained by removing N-1 links from a network (the number of links in a minimally connected network) and then calculating the link to node ratio. Combat networks should have a neutrality rating of between 0.8 and 1.2.

Clustering Coefficient. Clustering is one way to measure cohesion in a network. The clustering coefficient is the fraction of node triples in a network that have their third edge filled to complete a triangle. If a network has a high clustering coefficient then there is tight cohesion in the network (many triangles). The overall clustering coefficient is calculated globally over an entire network but individual nodes can have a clustering measurement as well. Skew-degree distribution networks have good clustering properties in the localities of the largest hubs but low clustering away from hubs of high activity. This local clustering provides the type of cohesion and mutual support that military operations require. For this reason, combat networks, like other adaptive networks, should have a skew distribution of clustering coefficients as well.[14]

Robustness. Robustness measures the extent to which a network can avoid catastrophic failure as links or nodes are removed. Robustness is usually determined by analyzing how network properties such as the size of the giant component, characteristic path lengths or betweenness, change with removal of nodes or links. The opposite of a robust network is a brittle network.[15] Figure II.10 shows the size of the giant component in a military e-mail network with skew degree distribution plotted against the number of nodes removed from the network. Random removal results in an almost horizontal line (meaning the giant component stays connected), whereas removal by degree rank (highest first) shows a rapid disconnection. This is intuitive because there are many more low-degree nodes in a complex network than high-degree nodes, so a random selection of nodes should favor low-degree nodes over those with high degree. Combat networks should therefore be extremely resilient to random attack but can be very susceptible to focused attack. It is therefore extremely important to deny knowledge of detailed network structure to adversaries.

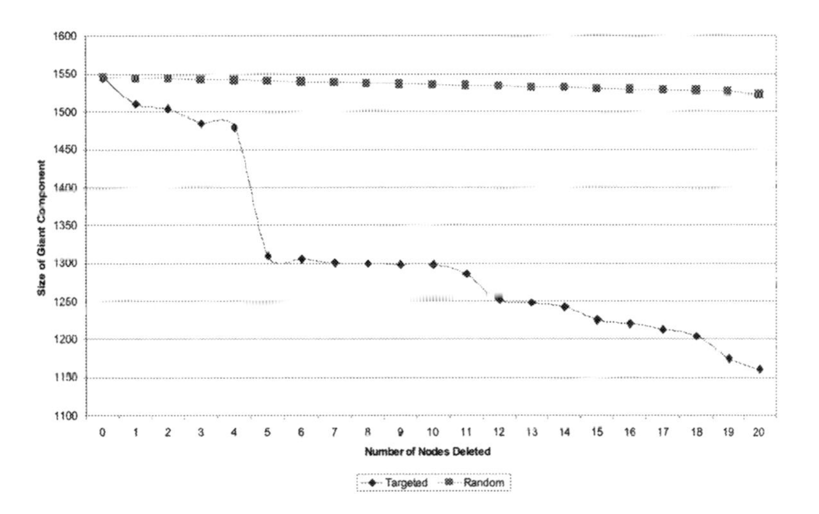

Figure II.10—Robustness in a Complex Network

Diffusion Rates. A network's diffusion rate is average number of nodes reachable by traversing exactly l links. Figure II.11 compares the diffusion rates of all the networks listed in Table II.1. The fastest rate, of course, is found in the maximally connected network because each node is directly connected to every other node. The slowest rate occurs in the minimally connected network, since this network contains no shortcuts. Between these upper and lower bounds is the rudimentary shape of diffusion patterns typical of complex networks (the curves in Figure II.11 would be more pronounced if the networks had more links and nodes). Clustering, path lengths, degree distribution, betweenness and neutrality all contribute to the shape of these curves. Figure II.12 shows how a 400-node network would have different diffusion curves for different connection probabilities (a proxy for link density). The curves get steeper as connection probabilities increase. Figure II.13 shows that reducing the number of nodes by an order of magnitude requires an increase in the connection probability by an order of magnitude to achieve the same diffusion rates. Figure II.14 shows how more nodes increase the diffusion rate for the same connection probability. All of these diffusion curves show how the increasing the number of links or nodes has a non-linear effect on network diffusion rates.

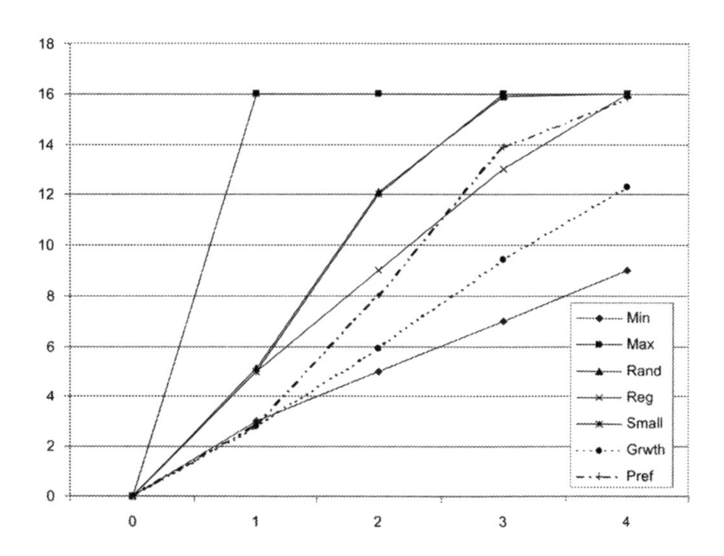

Figure II.11—Generic Diffusion Profiles: Selected Networks

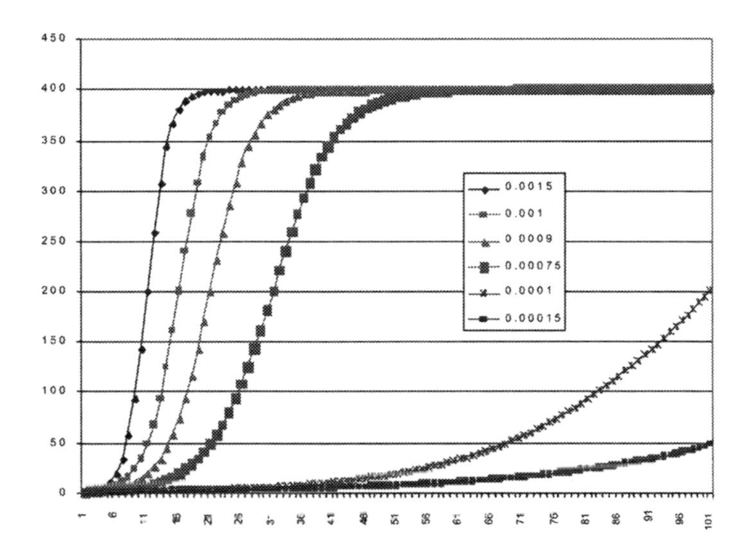

Figure II.12—Diffusion Rates, 400 Node Nerwork

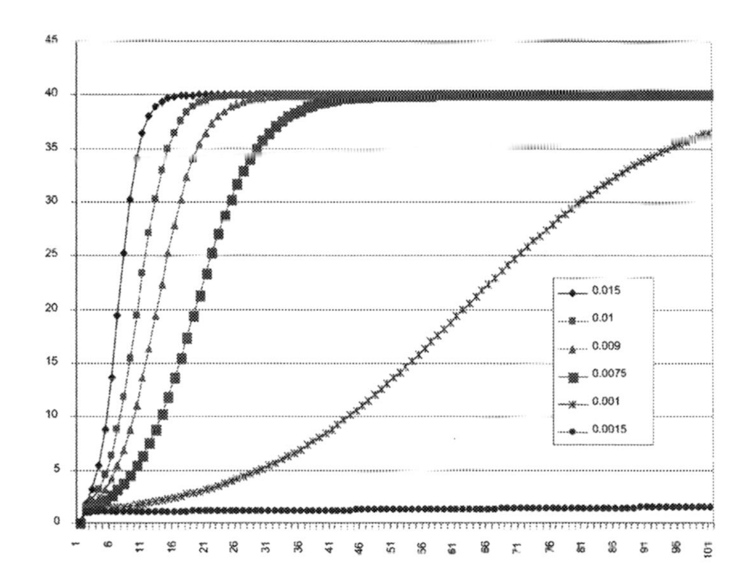

Figure II.13—Diffusion Rates, 40 Node Network

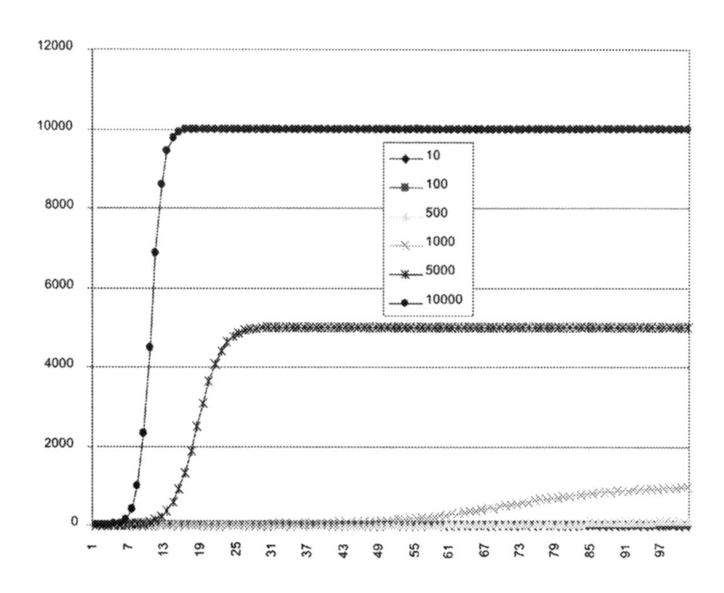

Figure II.14—Diffusion Rates, P(Connection) = 0.001

[1] CPL is computed as follows. First, find and measure the shortest paths from Node 1 to Nodes 2 through N, the shortest paths from Node 2 to Node 1 and to Nodes 3 through N, etc. Next, compute the average measurement for each node, and then rank these averages. CPL is the middle-ranked value (for an odd number of nodes) or the average of the two middle-ranked values (for an even number of nodes).

[2] Watts, Duncan, *Small Worlds*, (Princeton University Press, New York, 1999).

[3] Watts, Small Worlds.

[4] Barabasi, *Linked: The New Science of Networks*, Chapter 6.

[5] $n! = n(n-1)(n-2)(n-3)\ldots(1)$. n! (spoken, "n factorial") is one of the highest level of "computational complexity" in network mathematics.

[6] To be technically accurate, this network is actually pseudo-random, since it was created with a computer and true random sequences cannot be guaranteed by computer algorithm.

[7] Barabasi, Chapter 11.

[8] Barabasi, Chapter 6.

[9] See Oz Shy, *The Economics of Networked Industries*, (Cambridge University Press, New York, 2001), and Barabasi, Chapter 7.

[10] A Power Law is an equation of the form $P[x = X] \sim x^{-a}$. To more fully appreciate the behavior of these functions, the reader is encouraged to experiment with the Power Law using easily available software like MicroSoft Excel™ and sample values of x and a.

[11] Stefan Wuchty and Peter F. Stadler, "Centers of Complex Networks," Santa Fe Institute Working Paper, 2002-09-052, September 2002.

[12] Sergi Valverde and Ricard V. Solé, "Internet's Critical Path Horizon," Santa Fe Institute Working Paper, 2004-06-010, June 2004.

[13] M. E. J. Newman, "The Structure and Function of Complex Networks," SIAM Review 45, 167-256 (2003).

[14] Newman, M. E. J., Strogatz, S. H., and Watts, D. J., "Random Graphs with Arbitrary Degree Distributions and their Applications," Phys. Rev. E, 64, 026118 (2001).

[15] See http://www.santafe.edu/sfi/research/focus/robustness/index.html, accessed 11 Oct 2002, for a deeper technical treatment of robustness.

Appendix III: Bibliography

The major ideas in this book are drawn from a diverse array of technical research, including recent research into the science of complex networks, diffusion models, social network analysis, multi-scale representations, biological adaptation, complex control theory and the physics of information. Although the audience for this book—military officers and defense community professionals—are quite well educated, almost none of this research is common to the typical curricula of military education or business schools. To encourage further investigation by interested readers, this bibliography lists the top 100 reference documents for distributed networked operations research.

Arthur, W. Brian. Increasing Returns and the Two Worlds of Business. Santa Fe Institute Working Paper, no. 96-05-028. Santa Fe: Santa Fe Institute, 1996.

Ashby, Ross W. Introduction to Cybernetics, Part II. New York: Chapman & Hall, Ltd., 1957.

Axelrod, Robert. Advancing the Art of Simulation in the Social Sciences. Santa Fe Institute Working Paper, no. 97-04-048. Santa Fe: Santa Fe Institute, 1997.

________. The Complexity of Cooperation: Agent-Based Models of Competition and Collaboration. Princeton: Princeton University Press, 1997.

Axtell, Robert. The Emergence of Firms in a Population of Agents. Santa Fe Institute Working Paper, no. 99-03-019. Santa Fe: Santa Fe Institute, 1999.

Barabási, Albert-László. Linked: How Everything Is Connected to Everything Else and What It Means. New York: Plume, 2003.

Barwise, Jon and Jerry Seligman. Information Flow: The Logic of Distributed Systems. New York: Cambridge University Press, 1997.

Bar-Yam, Yaneer. Dynamics of Complex Systems. Reading, MA: Addison-Wesley, 1997.

________. Multiscale Analysis of Littoral Warfare. CNO Strategic Studies Technical Paper, 2002.

________. Making Things Work: Solving Complex Problems in a Complex World. Cambridge, MA: NECSI Knowledge Press, 2004.

Beyerchen, A. "Clausewitz, Nonlinearity, and the Unpredictability of War," International Security, 17 (Winter 1992/93): 59-90.

Bolmarchic, Joe. "Who Shoots How Many," Lecture. WINFORMS Conference, Washington, DC: 22 January 2003.

Bonabeau, Eric, Marco Dorigo and Guy Theraulaz. Swarm Intelligence: From Natural to Artificial Systems (Santa Fe Institute Studies in the Sciences of Complexity Proceedings) New York: Oxford University Press, 1999.

Bonabeau, Eric, Laurent Dagorn, and Pierre Freon. Scaling in Animal Group-Size Distributions. Santa Fe Institute Working Paper, no. 99-01-005. Santa Fe: Santa Fe Institute, 1999.

Bracken, J., M. Kress, and R.E. Rosenthal. Warfare Modeling. Danvers, MA: Wiley and Sons, 1995.

Braitenberg, Valentino. Vehicles. Cambridge, MA: MIT Press, 1998.

Camacho, Juan and Ricard V. Sole. Scaling and Zipf's Law in Ecological Size Spectra. Santa Fe Institute Working Paper, no. 99-12-076. Santa Fe: Santa Fe Institute, 1999.

Camazine, Scott, Jean-Louis Deneubourg, Nigel R. Franks, James Sneyd, Guy Theraulaz, and Eric Bonabeau. Self-Organization in Biological Systems. Princeton: Princeton University Press, 2001.

Cares, Jeffrey R. and John Q. Dickmann. "Information Age Warfare Sciences." Lecture. Preserving National Security in a Complex World Conference. Cambridge, MA: 12-14 December 1999.

________. "An Information Age Combat Model". Washington DC: unpublished DoD White Paper, 2005. <http://www.dnosim.com/papers>

________. Rule Sets For Sense And Respond Logistics: The Logic Of Demand Networks. Newport, RI: Unpublished DoD white paper, 2004. <http:// www.dnosim.com/papers>

Cares, Jeffrey. "The Fundamentals of Salvo Warfare." Masters Thesis, Naval Postgraduate School, Monterey, CA: 1990.

Cares, Jeffrey, and CAPT Linda Lewandowski, USN, Sense and Respond Logistics: The Logic of Demand Networks. Washington, DC: Unpublished DoD white paper, 2002. <http://www.dnosim.com/papers>

Cares, Jeffrey, Raymond A. Christian, and Robert C. Manke. Fundamentals of Distributed, Networked Forces and the Engineering of Distributed Systems. NUWC-NPT Technical Report 11,366. Newport, RI: Unpublished. 2002. <http:// www.dnosim.com/papers>

Chaitin, Gregory J. The Limits of Mathematics: A Course on Information Theory and the Limits of Formal Reasoning. Singapore: Springer-Verlag Singapore Pte. Ltd., 1998.

Crutchfield, James P. and Peter Schuster. Evolutionary Dynamics. New York: Oxford University Press, 2003.

Epstein, Joshua M. Nonlinear Dynamics, Mathematical Biology, and Social Science. Reading, MA: Addison-Wesley, 1997.

Gell-Mann, Murray. The Quark and the Jaguar: Adventures in the Simple and the Complex. New York: W.H. Freeman and Company, 1994.

Gell-Mann, Murray and Seth Lloyd. Effective Complexity. Santa Fe Institute Working Paper, no. 03-12-068. Santa Fe: Santa Fe Institute, 2003.

Gleiss, Petra M., Peter F. Stadler, Andreas Wagner, and David A. Fell. Small Cycles in Small Worlds. Santa Fe Institute Working Paper, no. 00-10-058. Santa Fe: Santa Fe Institute, 2000.

Guerin, S. and Kunkle, D. "Emergence of constraint in self-organizing systems." Journal of Nonlinear Dynamics, Psychology, and Life Sciences 8 (April, 2004): 131-146.

Gambhir, M., Guerin, S., Kunkle, D., Harris, R. "Measures of Work in Artificial Life." 2004. <http://www.redfish.c om/research/MeasuresOfWorkInALife_v1_1.pdf>.

Ho, K.J., "An Analysis of the Characteristics of a Distributed System," Masters Thesis, Naval Postgraduate School, Monterey, CA: 2001.

Holland, J. H. Adaptation in Natural and Artificial Systems: An Introductory Analysis with Applications to Biology, Control and Artificial Intelligence, Second Edition, Cambridge, MA: MIT Press, 1992.

________. Hidden Order: How Adaptation Builds Complexity, Reading, MA: Addison-Wesley Publishing Company, 1995.

Hong, Lu and Scott E. Page. Diversity and Optimality. Santa Fe Institute Working Paper, no. 98-08-077. Santa Fe: Santa Fe Institute, 1998.

Horne, Gary and Sarah Johnson. Maneuver Warfare Science 2002. Quantico: US Marine Corps Project Albert, 2002.

________. Maneuver Warfare Science 2003. Quantico: US Marine Corps Project Albert, 2003.

Hughes, Wayne P., editor. Military Modeling. Alexandria, VA: Military Operations Research Society: 1989.

Hut, Piet, David Ruelle, and Joseph F. Traub. Varieties of Limits to Scientific Knowledge. Santa Fe Institute Working Paper, no. 98-02-015. Santa Fe: Santa Fe Institute, 1998.

Ilachinski, A. Center for Naval Analyses. Land Warfare and Complexity, Part I: Mathematical Background and Technical Sourcebook. CIM-461. Alexandria, VA: July 1996.

________. Land Warfare and Complexity, Part II: An Assessment of the Applicability of Nonlinear Dynamics and Complex Systems Theory to the Study of Land Warfare. CRM-68. Alexandria, VA: July 1996.

Jain, Sanjay and K. Sandeep. "Graph Theory and the Evolution of Autocatalytic Networks." Nonlinear Sciences. 30 October 2002. <http://arxiv.org/abs/nlin.AO/021007 0>.

________. Crashes, Recoveries, and "Core-Shifts" in a Model of Evolving Networks. Santa Fe Institute Working Paper, no. 01-12-075. Santa Fe: Santa Fe Institute, 2001.

Jarvis, David A. "A Methodology For Analyzing Complex Military Command And Control (C2) Networks." Newport RI: unpublished DoD White Paper, 2005. <http://www.dnosim.com/papers>

Jen, Erica. Stable or Robust? What's the Difference? Santa Fe Institute Working Paper, no. 02-12-069. Santa Fe: Santa Fe Institute, 2002.

Jost, Juergen. External and Internal Complexity of Complex Adaptive Systems. Santa Fe Institute Working Paper, no. 03-12-070. Santa Fe: Santa Fe Institute, 2003.

Kauffman, Stuart A. The Origins of Order. New York: Oxford University Press, 1993.

________. At Home in the Universe. New York: Oxford University Press, 1995.

Kauffman, Stuart A. and William G. Macready. Technological Evolution and Adaptive Organizations. Santa Fe Institute Working Paper, no. 95-02-008. Santa Fe: Santa Fe Institute, 1995.

Kauffman, Stuart A., Jose Lobo, and William G. Macready. Optimal Search on a Technology Landscape. Santa Fe Institute Working Paper, no. 98-10-091. Santa Fe: Santa Fe Institute, 1998.

Kohler, Timothy A. and George J. Gumerman. Dynamics in Human and Primate Societies. New York: Oxford University Press, 1999.

Lloyd, Seth. "Learning How to Control Complex Systems." Spring 1995. <http://www.santafe.
edu/sfi/publication s/Bulletins/bulletin-spr95/10control.html>.

Lobo, Jose and William G. Macready. Landscapes: A Natural Extension of Search Theory. Santa Fe Institute Working Paper, no. 99-05-037. Santa Fe: Santa Fe Institute, 1999.

Mackay, Charles. Great Popular Delusions and the Madness Of Crowds. London: L. C. Page & Company, 1932.

McCloud, Scott. Understanding Comics. New York: Harper Collins books, 1994.

Moore, Cristopher and M. E. J. Newman. Epidemics and Percolation in Small-World Networks. Santa Fe Institute Working Paper, no. 00-01-002. Santa Fe: Santa Fe Institute, 2000.

Newman, M. E. J. Clustering and Preferential Attachment in Growing Networks. Santa Fe Institute Working Paper, no. 01-03-021. Santa Fe: Santa Fe Institute, 2001.

________. Random Graphs as Models of Networks. Santa Fe Institute Working Paper, no. 02-02-005. Santa Fe: Santa Fe Institute, 2002.

________. The Spread of Epidemic Disease on Networks. Santa Fe Institute Working Paper, no. 02-04-20. Santa Fe: Santa Fe Institute, 2002.

________. "The Structure and Function of Complex Networks." SIAM Review, 45 (2003): 167-256.

________. Who is the Best Connected Scientist? A Study of Scientific Coauthorship Networks. Santa Fe Institute Working Paper, no. 00-12-064. Santa Fe: Santa Fe Institute, 2000.

Newman, M. E. J. and R. G. Palmer. Modeling Extinction. New York: Oxford University Press, 2003.

Newman, M. E. J., Emily M. Jin, and Michelle Girvan. The Structure of Growing Social Networks. Santa Fe Institute Working Paper, no. 01-06-032. Santa Fe: Santa Fe Institute, 2001.

Newman, M. E. J., Michelle Girvan, and J. Doyne Farmer. Optimal Design, Robustness, and Risk Aversion. Santa Fe Institute Working Paper, 02-02-009. Santa Fe: Santa Fe Institute, 2002.

Newman, M. E. J., S. H. Strogatz, and D. J. Watts. Random Graphs with Arbitrary Degree Distribution and Their Applications. Santa Fe Institute Working Paper, no. 00-07-042. Santa Fe: Santa Fe Institute, 2000.

Norretranders, Tor. The User Illusion. New York: Penguin Books, 1998.

Page, Scott E. Uncertainty, Difficulty, and Complexity. Santa Fe Institute Working Paper, no. 98-08-076. Santa Fe: Santa Fe Institute, 1998.

Reidys, Christian, Christian V. Forst, and Peter Schuster. Replication and Mutation on Neutral Networks: Updated Version 2000. Santa Fe Institute Working Paper, no. 00-11-061. Santa Fe: Santa Fe Institute, 2000.

Richards, Diana, Whitman A. Richards, and Brendan D. McKay. Collective Choice and Mutual Knowledge Structures. Santa Fe Institute Working Paper, no. 98-04-032. Santa Fe: Santa Fe Institute, 1998.

Rosen, Robert. "Anticipatory Systems: Philosophical, Mathematical and Methodological Foundations." IFSR International Series on Systems Science & Engineering. 1 (1985). 45-220.

Shalizi, Cosma Rohilla and James P. Crutchfield. Computational Mechanics: Pattern and Prediction, Structure and Simplicity. Santa Fe Institute Working Paper, no. 99-07-044. Santa Fe: Santa Fe Institute, 1999.

________. Pattern Discovery and Computational Mechanics. Santa Fe Institute Working Paper, no. 00-01-008. Santa Fe: Santa Fe Institute, 2000.

Shannon, Claude E. "A Mathematical Theory of Communication." The Bell System Technical Journal, Vol. 27, January 2005: 379-423, 623-656.

Shubik, Martin. Game Theory, Complexity, and Simplicity Part I: A Tutorial. Santa Fe Institute Working Paper, no. 98-04-027. Santa Fe: Santa Fe Institute, 1998.

________. Game Theory, Complexity, and Simplicity Part II: Problems and Applications. Santa Fe Institute Working Paper, no. 98-04-028. Santa Fe: Santa Fe Institute, 1998.

________. Game Theory, Complexity, and Simplicity Part III: Critique and Prospective. Santa Fe Institute Working Paper, no. 98-04-029. Santa Fe: Santa Fe Institute, 1998.

Shy, Oz. The Economics of Network Industries. New York: Cambridge University Press, 2001.

Sole, Ricard V. and Romualdo Pastor-Satorras. Complex Networks in Genomics and Proteomics. Santa Fe Institute Working Paper, no. 02-06-026. Santa Fe: Santa Fe Institute, 2002.

Sole, Ricard V. and M. E. J. Newman. Patterns of Extinction and Biodiversity in the Fossil Record. Santa Fe Institute Working Paper, no. 99-12-079. Santa Fe: Santa Fe Institute, 1999.

Sole, Ricard V., Ramon Ferrer, Isabel Gonzalez-Garcia, Josep Quer, and Esteban Domingo. Red Queen Dynamics, Competition, and Critical Points in a Model of RNA Virus Quasispecies. Santa Fe Institute Working Paper, no. 97-11-085. Santa Fe: Santa Fe Institute, 1997.

Stephens, Christopher R. and Peter F. Stadler. Landscapes and Effective Fitness. Santa Fe Institute Working Paper, no. 02-11-061. Santa Fe: Santa Fe Institute, 2002.

Strogatz, Steven. Sync: The Emerging Science of Spontaneous Order. New York: Hyperion Books, 2003.

Traub, J. F. and A. G. Werschulz. Complexity and Information. New York: Cambridge University Press, 1998.

Traub, Joseph F. On Reality and Models. Santa Fe Institute Working Paper, no. 96-03-010. Santa Fe: Santa Fe Institute, 1996.

Valverde, Sergi and Richard V. Sole. Self Organized Critical Traffic in Parallel Computer Networks. Santa Fe Institute Working Paper, no. 01-11-071. Santa Fe: Santa Fe Institute, 2001.

________. Hierarchical Small Worlds in Software Architecture. Santa Fe Institute Working Paper, no. 03-07-044. Santa Fe: Santa Fe Institute, 2003.

________. Internet's Critical Path Horizon. Santa Fe Institute Working Paper, no. 04-06-010. Santa Fe: Santa Fe Institute, 2004.

Valverde, Sergi, Ramon Ferrer Cancho, Richard V. Sole, and Jose M. Montoya. Selection, Tinkering, and Emergence in Complex Networks. Santa Fe Institute Working Paper, no. 02-07-029. Santa Fe: Santa Fe Institute, 2002.

Wagner, Andreas. Causality in Complex Systems. Santa Fe Institute Working Paper, no. 97-08-075. Santa Fe: Santa Fe Institute, 1997.

________. Robustness, Evolvability, and Neutrality. Santa Fe Institute Working Paper, no. 04-12-030. Santa Fe: Santa Fe Institute, 2004.

Wasserman, Stanley and Katherine Faust. Social Network Analysis. New York: Cambridge University Press, 1994.

Watts, Barry D. Institute for National Strategic Studies, National Defense University. McNair Paper Number 52: Clausewitzian Friction and Future War. Washington D.C.: October 1996.

Watts, D. J. Six Degrees: The Science of a Connected Age. New York: Norton, 2003.

________. Small Worlds: The Dynamics of Networks Between Order and Randomness. Princeton, NJ: Princeton University Press, 1999.

West, Geoffrey B., James H. Brown, and Brian J. Enquist. A General Model for the Origin of Allometric Scaling Laws in Biology. Santa Fe Institute Working Paper, no. 97-03-019. Santa Fe: Santa Fe Institute, 1997.

Wolpert, David H. and William G. Macready. No Free Lunch Theorems for Search. Santa Fe Institute Working Paper, no. 95-02-010. Santa Fe: Santa Fe Institute, 1995.

________. Self-Dissimilarity: An Empirical Measure of Complexity. Santa Fe Institute Working Paper, no. 97-12-087. Santa Fe: Santa Fe Institute, 1997.

Wutchy, Stefan and Oeter F. Stadler. Centers of Complex Networks. Santa Fe Institute Working Paper, no. 02-09-052. Santa Fe: Santa Fe Institute, 2002.

Young, H. Peyton. The Diffusion of Innovations in Social Networks. Santa Fe Institute Working Paper, no. 02-04-018. Santa Fe: Santa Fe Institute, 2002.

Zurek, W. H. Complexity, Entropy and the Physics of Information. New York: Addison-Wesley, 1991.

About the Author

Jeff Cares is one of the top thought-leaders in Information Age innovation and has been featured in such Information Age bellwethers as Wired and Fast Company. He has published pioneering work in the application of New Science techniques to military problems and lectures internationally on the future of military forces.

Index

D

Z

978-0-595-37800-5
0-595-37800-5

Printed in the United States
52697LVS00003B/140